Maize for the Gods

The publisher gratefully acknowledges the generous support of the Ahmanson Foundation Humanities Endowment Fund of the University of California Press Foundation.

Maize for the Gods

*Unearthing the 9,000-Year
History of Corn*

Michael Blake

UNIVERSITY OF CALIFORNIA PRESS

University of California Press, one of the most
distinguished university presses in the United States,
enriches lives around the world by advancing scholarship
in the humanities, social sciences, and natural sciences. Its
activities are supported by the UC Press Foundation and
by philanthropic contributions from individuals and
institutions. For more information, visit www.ucpress.edu.

University of California Press
Oakland, California

Library of Congress Cataloging-in-Publication Data

Blake, Michael, 1953– author.
    Maize for the Gods : unearthing the 9,000-year history
of corn / Michael Blake.
        pages   cm
    Includes bibliographical references and index.
    ISBN 978-0-520-27687-1 (cloth : alk. paper) —
    ISBN 978-0-520-28696-2 (pbk. : alk. paper) —
    ISBN 978-0-520-96169-2 (ebook)
    1. Corn—History. I. Title. II. Title: Unearthing the
9,000-year history of corn.
    SB191.M2B657   2015
    633.1'5—dc23   2015004625

24   23   22   21   20   19   18   17   16   15
10   9   8   7   6   5   4   3   2   1

Contents

Illustrations

MAPS

Acknowledgments

I conceived of this book as a way to make sense of the vast amount of new archaeological research that was appearing on the history of food and agriculture and, in particular, the history of maize—one of the world's most successful food crops. So many different lines of evidence were expanding and changing my understanding of the complex and ancient relationships between people and maize that I thought it would be helpful to pull some of the threads together in one place. This idea was sparked by a two-day symposium, held at the Society for American Archaeology meetings in Montreal in 2004, that I co-organized with John Staller and John Hart. The symposium resulted in a massive collection of scientific papers that were published in 2006 in *The Histories of Maize,* edited by John Staller, Bruce Benz, and Robert Tykot. Building on this publication and the work of a host of other scholars, I started to craft a concise snapshot of what we know about the archaeology, dating, history, genetics, chemistry, technology, and symbolism of maize's origins—from an archaeologist's perspective. What I imagined might take a few months or a year to accomplish wound up taking several years, and even now I feel like I've only just scratched the surface of this fascinating history.

This book could never have come to fruition without the endless patience and encouragement of Blake Edgar, senior editor at the University of California Press. The extent to which this book is readable is the result of Blake's clear vision and insightful suggestions. I am grateful to both Merrik Bush-Pirkle and Dore Brown for shepherding the project through the exacting final stages of production, and I thank Genevieve Thurston for her superb copyediting.

My preoccupation with the origins and history of maize agriculture has grown out of many decades of research and conversations with a host of colleagues, students, and friends who are too numerous to mention individually. I would be remiss, however, if I didn't acknowledge the generous encouragement, help, and contributions of the following people. Bruce Benz, John Staller, R. G. Matson, Bruce Miller, Gary Coupland, Dana Lepofsky, Margaret Nelson, Ben Nelson, John Hart, and Hannes Dempewolf all provided invaluable intellectual support during the manuscript preparation by reading drafts of chapters or simply discussing ideas about the anthropology and archaeology of agriculture and corn. I thank the two reviewers—Karl Taube and Bob Hard—who generously gave of their time to make many helpful suggestions for improving the manuscript. The ideas and themes that I explore in this volume are the direct result of conversations I had with colleagues while excavating ancient village sites, working in the lab, writing publications and reports, and planning research. I am especially grateful to my friend and colleague John Clark and to the late Tom Lee Jr. and Gareth Lowe of the New World Archaeological Foundation-Brigham Young University, both of whom have generously supported the work John and I carried out in Chiapas over many years. I thank Barbara Voorhies, Brian Hayden, Vicki Feddema, Richard Lesure, David Cheetham, Rob Rosenswig, Brian Chisholm, Mike Richards, Diana Moreiras, Ken Hirth, David Webster, Lindi Masur, John Smalley, Cecilia Canal, Nadine Gray, Virginia Popper, Hendrik Poinar, John Hart, Hugh Iltis, Robert Bird, Michael Moseley, Richard Burger, Ruth Shady, Mary Pohl, Sonia Zarrillo, Jonathan Haas, Steven LeBlanc, Jane and David Kelley, Kevin Hanselka, Scott Raymond, Gary Crawford, and Mario Rivera for generously sharing their ideas and data.

The maps in this volume are the result of the creative and technical skills of Nick Waber, who never failed to solve a mapping problem with speed and wit. Many are based on data gathered and produced as part of the *Ancient Maize Map,* an online database that tracks the dated samples of maize throughout the Americas. I thank all my generous colleagues in a dozen different countries who have contributed to this database over the past eight years and am enormously grateful to the mapping team that helped build it: Bruce Benz, Sue Formosa, Kisha Supernant, Alex Wong, Diana Moreiras, and the amazing duo of Nick Jakobsen and Ryan Wallace, the incomparable programmers from Culture Code. For their contributions of photos and drawings to help illustrate this volume, I am especially grateful to Kent Flannery, Joyce Marcus, John Doebley, Tony Ranere, Ken Hirth, Ryan Williams, Paul Minnis, Chris Hastorf, Fred McColly, John Doebley, Bruce Benz, David Cheetham, Karl Taube, Hector Neff, Sonia Zarrillo, and Cory Kratz.

Research support for my work has come from the Department of Anthropology and the Laboratory of Archaeology at the University of British Columbia. My thanks to former and current department heads David Pokotylo, John Barker, Susan Rowley, and Patrick Moore for their help in lightening the administrative load and to Patricia Ormerod, the manager of the Laboratory of Archaeology, for her constant assistance with collections and photography. I am grateful to the Social Science and Humanities Research Council of Canada for its generous funding of my archaeological fieldwork and research in Mexico and my subsequent research on the history of maize throughout the Americas.

My final and most heartfelt thanks goes to my wife, Susan, and our sons, David and D'Arcy. Their patience, encouragement, and good humor during this prolonged process has allowed them to reap what I have sown; they now know more about corn—much more—than they could ever have wished or imagined.

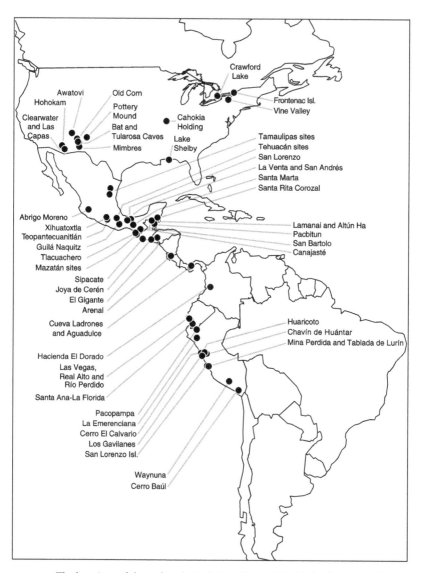

Crawford
Lake

Awatovi
Hohokam

Old Corn

Frontenac Isl.
Vine Valley

Clearwater
and Las
Capas

Pottery
Mound

Cahokia
Holding

Bat and
Tularosa Caves

Lake
Shelby

Mimbres

Tamaulipas sites
Tehuacán sites
San Lorenzo
La Venta and San Andrés
Santa Marta
Santa Rita Corozal

Abrigo Moreno
Xihuatoxtla
Teopantecuanitlán
Guilá Naquitz
Tlacuachero
Mazatán sites

Lamanai and Altún Ha
Pacbitun
San Bartolo
Canajasté

Sipacate
Joya de Cerén
El Gigante
Arenal

Cueva Ladrones
and Aguadulce

Huaricoto
Chavín de Huántar
Mina Perdida and Tablada de Lurín

Hacienda El Dorado

Las Vegas,
Real Alto and
Río Perdido

Santa Ana-La Florida

Pacopampa
La Emerenciana
Cerro El Calvario
Los Gavilanes
San Lorenzo Isl.

Waynuna
Cerro Baúl

MAP 0.1. The locations of the archaeological sites discussed in this book. (By Michael Blake and Nick Waber)

Introduction

Hardly a week goes by without a news report proclaiming a food production crisis in some region of the world. Both local and international headlines regularly catch one's eye with concerns about crops and food sources being impacted by global warming, energy production, genetic modification, regional droughts, plagues, contamination, and a complex tangle of other factors. Maize (*Zea mays* ssp. *mays* L.) is one of the three largest crops produced in the world today (along with wheat and rice) and is often at the forefront of these alarming reports: "Demand for ethanol spurs tortilla crisis in Mexico," "EU considers ordering France to lift GMO maize ban," "South Africa exports to Mexico to produce maize deficit."[1] The Mexican "tortilla crisis" of a few years ago has now been overshadowed by the country's drug wars, but both are persistent crises that show no signs of going away. They are the result of ongoing economic imbalances and the changing global demand for and flow of commodities, such as food, energy, and drugs, to name but a few.

It seems surprising that maize—a plant originally domesticated in Mexico some nine millennia ago—has become such an important commodity on the global market, in part because of its value for producing ethanol as a gasoline replacement, that it is now too expensive for tens of thousands of Mexicans who depend on it for their daily food. One of the ironies of this development is that maize may originally have been domesticated not only for food but also for its sugary juice, which can be fermented into alcohol. If this is true, then the significance of this parallel between the modern and ancient uses of maize lies in its diverse range of functions, symbolized in part by its use both as a food and as a source of alcohol. It may be difficult for most present-day maize consumers to appreciate why this is important to consider—but I think it helps us to understand that the food production and distribution problems and successes of today have, quite literally, deep roots in the past. Why is maize one of the three biggest world crops today and how did it get to be that way? I won't try to answer these questions in all their complexity in this book, but I will provide a concise exploration of maize's history from an archaeologist's perspective. Much of what we know of this plant's history comes from archaeological research—from the traces of maize recovered in ancient houses, refuse heaps, caves, and even lake bottoms. We find the stone tools for grinding maize, ceramic vessels for cooking and serving it and for brewing maize beer, storage containers for protecting the yearly harvest, paintings of maize on ancient building walls, depictions of it on pottery vessels, and so on. The history of maize is also hidden in its genetic code—and scientists are rapidly discovering more about the biological modifications that shaped the plant during its nine thousand or more years of entanglement with the humans who discovered it, cared for it, modified its structure, and were in turn transformed by their new botanical creation.

Before taking my first archaeology and geography courses at university (many, many years ago), I, like most people, had only the vaguest notion of where maize came from, what its wild relatives looked like, why they were domesticated, and how and why maize spread from its homeland in Mexico far and wide throughout the Americas and eventually around the globe. But through my own archaeological research over the past few decades, I've become increasingly fascinated with the plant and somewhat overwhelmed by the explosion of information about its history. Working with other archaeologists, farmers, plant biologists, geneticists, paleoethnobotanists, geographers, and many others who study the origins of agriculture and the evolution of human-plant relationships, I've also become increasingly aware that new and quite exciting understandings of maize's origins and histories are emerging.

There have been enormous advances in our knowledge of maize's history in the past two decades, just as there have been huge changes in the way maize has been incorporated into industrial agriculture, the global food industry, and now the energy sector. Many of the discoveries about how past peoples used and thought about maize and observations about how present-day people are transforming the plant suggest that much of maize's and maize users' histories are a long series of unintended consequences. One of these consequences may have been set in motion when the first harvesters of wild *Zea mays* decided to select and plant seeds from this tropical grass with sugary stalks, instigating the appearance of a productive seed head that we recognize as a miniature ear of maize with its characteristic cob. Similarly, present-day growers of maize who are producing the plant to create bio-fuel, in a laudable attempt to alleviate our dependence on petroleum, probably did not initially realize that one consequence of this innovation would be to deprive low-income families of an affordable food staple that has fed millions of people for centuries. But this, in all likelihood, is just the tip of the iceberg. Maize, as a global food source, both directly (for human consumers) and indirectly (for livestock), is produced in greater quantities than wheat and rice, and its shift to fuel use will have a ripple effect, and ultimately a tsunami impact, on all aspects of the global economy.[2] Similar shifts in maize's uses and importance as a food source took place in antiquity, and they are likely to happen again. As in the past, such a massive shift in agricultural production is likely to transform technologies, ideologies, social and political relations, food preferences, demographics, and health. And, just as in the past, it is likely to transform people's relationships with one another and with their gods.

YOUTHFUL ENCOUNTERS WITH CORN

Like millions of North Americans I grew up in an area where maize is called "corn,"[3] and my fondest memories of the food is as corn on the cob, consumed voraciously during summer picnics and barbeques. It was cheap, plentiful, and fresh—even around the Interior Plateau city of Kamloops, British Columbia, where I grew up—and was all the more special because it was only available for a short time during late summer. As a child I remember eating the kernels fresh and uncooked right off the cobs that I picked myself—ten cents a dozen—and I loved their juicy sweetness. The pallid, pasty canned corn, and sometimes creamed corn, that my mother served us during winter months was far less appetizing, and my siblings and I seldom ate it gracefully. Yet we eagerly devoured cartons of Kellogg's Corn Pops and other corn-based confections that passed for nutritious breakfast foods—just as millions of children do today. How can maize look and taste so different in its various forms?

Later in my youth I was in for an even bigger surprise about the ways in which maize could be consumed. As a teenager hitchhiking through East and Southern Africa during a gap year between high school and college, I one day found myself in the dusty main square of a small rural community in southern Kenya. After a while what appeared to be an ancient water tanker truck pulled up under the shade of a large tree. The driver and his helper hopped down out of the cab while people of all ages appeared with buckets, plastic bottles, and tin cans, crowding around the spigot protruding from the back of the truck. I was thirsty after the long trip to the village, so I asked the driver if I could have some water too. He laughed and said that he didn't have any water, but for a few cents I could fill up my canteen with maize beer. As an underage teen I was impressed—until I filled my bottle and took a swig. With dismay I realized that in texture and taste the beer was almost exactly like my mother's creamed corn—only with a slight fizz and somewhat sour taste. It was food and drink combined—but it was not quite to my taste. This type of locally made maize beer is still commonly produced and consumed throughout East Africa, and in many regions the drink, known as *busaa* in and around Nairobi and as *mayuek* among the Okiek people of southern Kenya, is an integral part of funerary rituals and initiation ceremonies (figure o.1).[4]

My next encounter with maize on the trip was equally revealing but much more agreeable. A kind driver picked me up on a remote stretch of highway in northern South Africa. He took me as far as his village and, as it was getting dark, invited me to stay with him and his family and share their dinner. I was both grateful and relieved; it had been a long day of traveling with no opportunity to eat. Soon after our arrival, my host's wife placed on the table an enormous bowl filled with a stiff, white porridge they called mealie-meal. That, along with some steamed greens, was our dinner. My new friend and his children laughed with pleasure as I devoured my bowl of mealie-meal—quite disbelieving my claim that I'd never eaten it before. It was delicious and filling and kept me going during my travels all the next day. My hosts explained that mealie-meal was coarse-ground white maize flour that was boiled up as a porridge and that it was the basic element of every meal. Some of the maize they grew themselves, but most of it was purchased already ground and ready to cook.

At the time it didn't occur to me that maize might not be native to Africa, just as I had assumed that maize had always grown north of the 49th parallel in Western Canada. It wasn't until college that I read several eye-opening pieces that put my intercontinental maize experiences in

FIGURE 0.1. Okiek men drinking maize beer—*mayuek*—during a girls' initiation ceremony in Sinentaaiik, Kenya, in April 1985. (Photograph courtesy of Corinne A. Kratz)

perspective. The first was a chapter in my human geography course textbook that described the history of maize and what was then known about its origins. The chapter was a reprint of a paper by botanist Paul Mangelsdorf, who wrote that the ancient Indians of Mexico had domesticated maize by taming a wild ancestor of the plant. The second was a chapter in my introductory textbook in New World archaeology by William Sanders and Joseph Marino that described archaeologist Richard "Scotty" MacNeish's (then) recent discoveries of Mexico's earliest maize in several Archaic period caves in the states of Tamaulipas and Puebla.[5] Both chapters fanned my interest in maize. Perhaps the most startling revelation was that domesticated maize had to be planted by humans; it could not self-propagate. This put into focus for me the importance of understanding how a wild plant (or animal for that matter) becomes domesticated in the first place and subsequently transformed to the point where it might actually fail to survive without human intervention. It also made it clear that my encounters with maize in the temperate regions of North America, where I grew up, and the subequatorial tropics of Africa, where I had backpacked, were the result of a long history of migration and travel, both of peoples and their resources, in ancient times. I wanted to learn more.

In 1974, after finishing my second year of college, I decided to take another gap year and travel some more. The annual meetings of the American Anthropological Association were scheduled to take place in Mexico City in November. Along with some classmates from Cariboo College (now Thompson Rivers University) in Kamloops I hitchhiked through the US Southwest and then into Mexico, eventually arriving in Mexico City. My traveling companions and I were able to attend presentations by a number of archaeologists and anthropologists working in Mexico, and we were very impressed by the anthropological luminaries of the time (especially Margaret Mead—although she wasn't much interested in maize). This began a seven-month trip through Mexico and many other countries in Latin America—one that set the course of my future studies in anthropology and, although I didn't know it at the time, my prolonged interest in maize.

Traveling on a shoestring budget required that we eat cheaply and stay in inexpensive hotels or, even better, camp in order to make our meager funds last. Eating as frugally as possible meant eating maize, and in Mexico that meant tortillas. We lined up every few days outside the local *tortillería* where, for a few pesos, we could buy more tortillas than we could eat before they turned stale and moldy. Tortillas fresh off the *tortillería*'s conveyer belt are exquisite, and, like mealie-meal or maize beer in Southern and East Africa, hominy in the American South, or creamed corn from a can in Kamloops (ca. 1965), they are filling, nutritious, and cheap. As we traveled farther south through Central America and eventually into South America, tortillas disappeared, and our maize consumption, along with that of the locals, changed dramatically.

In Peru, for example, we were treated to popcorn and *chicha*—South American maize beer. I remembered from my Sanders and Marino text that *chicha* was one of the most common forms of maize consumed in Inca times and earlier. Again the connection between maize and alcohol was both clear and ancient. Was this the case everywhere? Why was the link so clear in the Andean region of South America and not very apparent in Mexico, where we had just been? What were people eating instead of maize for their daily nutrition? In Ecuador and Colombia many foods were part of the mix, and the diversity was astonishing: roast guinea pigs, amaranth cakes, tubers of various kinds (including of course, potatoes), a wide array of tropical fruits, and many introductions from the Old World, both plants and domesticated animals. Maize was also present and important, but it didn't seem to be as pervasive as in Mexico.

FIGURE 0.2. Christine Hastorf demonstrating the water flotation method—used for the recovery of ancient plant remains—during visitor day at the Mimbres Archaeological Project in the mid-1970s. (Photograph courtesy of Paul Minnis)

A few years later, in 1976, I had the chance to participate in a long-term archaeological project in the Mimbres Valley of New Mexico directed by the archaeologist Steven LeBlanc. I had the good fortune of working closely with two budding paleoethnobotanists, Paul Minnis and Christine Hastorf (who were graduate students at that time and now are renowned experts in archaeobotany). I assisted them occasionally with water flotation of soil samples aimed at recovering ancient plant remains and also helped a little with the sampling and collecting of living plant specimens for comparative collections (figure 0.2). The first day that we spotted wood charcoal and the charred remains of maize and beans in the flotation samples was thrilling. Most of my previous archaeological work had been in British Columbia, where archaeologists were not yet floating samples to recover plant remains, and even if they had been doing so they would not have found maize and beans. Why was it so exciting to find charred maize and beans in the household refuse of the ancient Mimbreños? I'm not so sure, but it was probably the realization that these were domesticated plants that had either been brought by farmers as they migrated north from Mexico or traded and planted by neighboring peoples over the course of generations. Either way, this demonstrated the antiquity of both population movements and individual people's interactions over thousands of kilometers and hundreds or even thousands of years. For an archaeologist it is an exciting prospect to see and even touch humble remains that tell such a profound story.

A few years later, from 1977 to 1979, I participated as a research assistant in Brian Hayden's Coxoh Ethnoarchaeology project in several Maya communities in Highland Chiapas, Mexico, and western Guatemala. This gave my fellow project members and me a privileged opportunity to live with people whose daily routines revolved around maize. We witnessed, day after day, the cycle of human-maize interaction: clearing land and burning the forest, planting, weeding, harvesting, hauling, storing, husking, degraining, cooking, washing, grinding, cooking again, eating, drinking, discarding, trading, presenting offerings to the saints, feasting, and then planting again (figure 0.3). All of this maize-related activity was embedded in countless other webs of actions and relationships. Even so, in talking with Tzeltal Maya people in Chanal, Aguacatenango, and many neighboring communities in the highlands of Chiapas about their economic lives, maize always came out on top—it was the benchmark of productivity, independence, social standing, and connection to both their land and their ancestors. The people in these communities had always relied on and revered maize. They even believed that this was probably true for most people around the world. And they were correct. In fact, my friends in Chanal were relieved to know that my family and I also ate corn (and even grew it in our garden). They were quite shocked, however, that we didn't make it into tortillas—but still, we were seen as maize eaters.

My experiences doing ethnoarchaeological research in Maya communities in Chiapas led me to my doctoral dissertation research: the archaeological study of political and social organization at the Postclassic period Maya settlement of Canajasté, located on Chiapas's eastern border with Guatemala. Neither maize nor agriculture in general were the focus of my research, which I carried out in the early 1980s, but my excavations of household deposits and my explorations of one previously looted temple at the site turned up lots of evidence of maize and its importance in the Postclassic Maya economy. Using a flotation barrel designed and built by John Rick for Kent Flannery and Joyce Marcus during their excavations at San Jose Mogote, Oaxaca, which they had kindly passed on to me, their impoverished graduate student, I floated many samples of soil from the house middens. Very little actual maize

FIGURE 0.3. Household members discarding their daily trash and floor sweepings (which include maize cobs) in the maize, bean, and squash garden surrounding their home in Chanal, a Tzeltal town in highland Chiapas, Mexico, in 1977. (Photograph by Michael Blake)

could be found in most of the house refuse samples—all carefully scanned by Paul Minnis—but there were thousands of pottery fragments from wares used to cook, process, and serve maize and many fragments of the *manos* and *metates* used to grind maize. Most of these ancient belongings were identical to the tools I had seen and studied in Chanal and Aguacatenango in Chiapas, Mexico, and San Mateo Ixtatán in Guatemala, and I knew they were also present in hundreds of other communities throughout the Maya region and beyond. One of the most exciting discoveries, though, came when I was cleaning off the floor of the looted temple, Structure 9, right at the heart of the community. Hundreds of charred maize kernels, many of them fragmentary, had been left in a deposit on the temple floor.[6] Unfortunately, the looters' depredations had made it impossible to tell much about the deposit—but it may have been an offering of maize that had been burned on the temple floor, perhaps as part of a yearly rite related to the new planting or the harvest of maize (figure 0.4).

At about the same time that I was completing my doctoral research, John Clark and I began a long-term archaeological project on the Pacific coast of Chiapas that would become the basis of John's dissertation

FIGURE 0.4. Excavation crew member Vicente Pérez reconstructing the staircase of a temple platform (Structure 9) at the Postclassic period site of Canajasté (ca. 500–750 years old) in Chiapas, Mexico, in 1982. Burned maize kernels had been left as an offering on the floor inside the doorway at the top of the stairs. (Photograph by Michael Blake)

research.[7] This work was focused on the Early Formative period sites in the region around Mazatán, where a previous generation of archaeologists had discovered the well-preserved remains of some of Mesoamerica's earliest villages—dating between 3,500 and 3,000 years ago, some three millennia before the Postclassic site that I had studied on the other side of the Sierra Madre mountains that fringed the coast. Early in 1985 we began to find rich household refuse deposits, particularly at the sites of Aquiles Serdán, where Carlos Navarrete had worked in 1968, and Paso de la Amada, where Jorge Fausto Ceja Tenorio had worked in the 1970s.[8] Using Rick's flotation machine, which we had lugged with us, we water-screened and floated as many samples as we could. In the evenings, in our lab in Mazatán, we sorted through the float heavy fraction—sediments, charcoal, rootlets, tiny fragments of pottery, animal bones, and minute artifacts—in other words, all the materials that didn't float to the surface and so couldn't be scooped off with a sieve. In one of our first inspections of samples from Aquiles Serdán, we spotted among the fish bones and potsherd fragments some tiny charred maize kernels and cob fragments, a few charred beans, and some larger charred seeds that turned out to be avocado pits (figure 0.5). This was the ancient version of what we had just had for dinner.

FIGURE 0.5. Charred maize kernels (top) and cob fragments (bottom) recovered during the flotation of soil samples from Preclassic period house deposits (ca. 3,300 years old) at the site of Aquiles Serdán in the Mazatán region of Pacific coastal Chiapas, Mexico, 1985. (Photograph by Michael Blake)

Over several seasons of excavation in many different Early Formative villages, we frequently encountered buried refuse deposits that contained similar assemblages of macroremains, including maize. In a few samples we recovered fragments of small maize cobs. Vicki Feddema, one of my master's students at the University of British Columbia (UBC), took on the task of studying the paleobotanical remains from our excavations and described them for her master's thesis.[9] She concluded not only that maize was present in most of the samples that we took, but also that it was the most common of all the different species of identifiable plant remains, regardless of time period. But there was a nagging question— were these maize remains really as old as we thought they were? Could they have been intrusive? We thought it unlikely because most of the sites we were investigating had very few later deposits that could have been the source of the maize. John and Vicki and I selected a small number of charred maize kernels from well-dated, undisturbed layers at several sites and submitted them for radiocarbon dating using the **AMS (Accelerator Mass Spectrometry)** method. The samples returned dates with ages ranging between 3,500 and 3,000 years **cal BP** (**calibrated years before present**). John Clark reported these findings in his doctoral dissertation, and we also included them in a radiocarbon chronology of the Soconusco region—the Aztec name for this part of the Pacific coast.[10] We didn't know it at the time, but it turned out that these remains were, and still are as of this writing, among the earliest directly dated maize remains from Early Formative villages in Mesoamerica. There are earlier directly dated maize remains—some 3,000 years earlier—but these are mostly from pre-village, Archaic period societies. I expect that this will change as more samples are directly dated and as techniques of AMS dating and the range of materials that can be directly dated are improved.

One exciting avenue for studying ancient diet that was developed and came of age in the 1970s and 1980s is stable isotope analysis. Early in 1991 two of my colleagues at UBC, R. G. Matson and Brian Chisholm, had successfully looked at the stable carbon and nitrogen isotopes of human bone samples from Basketmaker II phase sites at Cedar Mesa, Utah. As others were doing throughout the Americas, they were able to use the results of their study to provide an independent line of evidence for the evaluation of maize's significance in the diet.[11] In an attempt to do the same thing for the Soconusco region of coastal Chiapas, Brian Chisholm and I, along with several other colleagues, measured the stable isotope values of a large sample of human bone remains spanning the entire sequence of human occupation of the region: from the Archaic to the Postclassic periods. One of our expectations was that we would see a significant, if not dramatic, increase in the stable carbon isotope ratio of our samples from the Archaic to the Formative periods, when, according to the paleobotanical evidence, maize started to be grown widely on the coast.[12] We found the opposite. The stable isotopes showed that stable carbon ratios were higher in the Archaic than in the Early Formative, and it wasn't until much later in the Formative that many of our samples started to show signs of increasing ratios. The chemical signatures suggested that, even though maize was physically present—and in fact ubiquitous—in Early Formative times, it might not have constituted the largest portion of the diet.

My worry, both at the time and now, was that my previous training and my experience of and exposure to maize and my understanding of the importance of maize for present-day peoples of Mesoamerica actively shaped my perceptions of how maize must have been used in the past. This worry made me start to question the assumption that clear evidence of how maize was used in its initial stages in one region (say the central Mexican highlands) was sufficient to explain how it was used in other regions (say the Pacific coast of Chiapas). How might the discrepancies between my expectations (based on a lifetime of experience, from eating corn-on-the-cob as a young lad in western Canada to discussing the economic importance of maize with several dozen Tzeltal farmers in the highlands of Chiapas) and the different lines of evidence (from stable carbon and nitrogen isotope analyses to high-precision AMS radiocarbon dating of three-and-a-half-thousand-year-old maize) lead to new ways of looking at the past? They might not provide immediate answers or resolve the contradictions and debates, but I think they could lead to interesting new perspectives and give rise to important new questions.

Now let's return to the connection between maize and alcohol. In 1995, while teaching an undergraduate course in Mesoamerican archaeology at UBC, one of my students—a mature student, shall we say—approached me about the topic he wished to take on for his essay assignment. For his paper, John Smalley wanted to combine his interests in Mesoamerican archaeology and ethnography with his hobby as a brewer—specifically, he wanted to look at the archaeological evidence for maize beer making in Mexico. He was excited about the ethnographies that he had recently read, which described the northern Mexican Tarahumara people's practice of brewing *chicha* using maize kernels and another type of beer using the juice squeezed from maize stalks. This was the first I'd heard about maize-stalk beer. As I read the drafts of John's paper and looked at some of the original sources, and as we discussed the intricacies of maize beer making, particularly maize-stalk beer making, I remembered seeing photos of chewed maize stalks in Scotty MacNeish's reports on his excavations in the Tehuacán caves. For both of us a question crystallized: could the Tarahumara practice of maize-stalk beer making be an ancient tradition, possibly one that goes back to the initial uses of maize's wild ancestor, teosinte?[13] As a first step to exploring this question, we decided to make some maize-stalk beer ourselves. We collected several large bags full of freshly cut corn stalks from a nearby farmer's field—just after the corn harvest was complete, of course—chopped and mashed them up, squeezed out the juice, and then set it aside to ferment. I won't say the resulting brew was delicious, but it was both refreshing and alcoholic (figure 0.6).

This question about the earliest uses of teosinte, along with my previous encounters with maize, both ancient and modern, have led me to an

FIGURE 0.6. John Smalley and Nadine Gray chopping up fresh maize stalks in order to extract their sugary juice, which will be used to make maize stalk beer, Vancouver, BC, Canada, 2007. (Photograph by Michael Blake)

expanding fascination with the new research about this crop—work that is taking place from Canada to Argentina and all points in between. I have been increasingly drawn into the vast literature dealing with maize archaeology, paleoethnobotany, microremains (such as pollen, phytoliths, and starch grains), genetics, linguistics, art, and symbolism in trying to answer rather specific questions. Could the earliest villagers on the coast of Chiapas have relied on maize as a staple? Were maize stalks used to make beer in ancient times, and might this have been one of the reasons for maize's early and rapid spread? Did maize farmers spread from region to region and expand at the expense of the earlier inhabitants, or did maize pass from region to region by trade, with people adopting maize for myriad reasons depending on how it fit into their existing economies? Many recent publications have observed that maize is one of our best known domesticated crops—and a small forest of trees has been converted into huge quantities of paper to publish this hard-won knowledge. This book adds to that pile of paper in an attempt to draw together my own views of this work and lay down a few stepping-stones on the path to formulating new research questions about maize's origins.

YESTERDAY'S NEWS

Not surprisingly, large oil companies are not so keen on supplanting gasoline with ethanol—whether it's made from maize or any other plant. A headline from Canada's *Globe and Mail* newspaper on January 30, 2007, announced: "Bush's ethanol plan getting cold shoulder: Exxon not onside; producer stocks founder." Still, there are many supporters of this trend—including companies and governments that see a bright future in generating energy from renewable resources. Perhaps ethanol will continue to be made from maize, but other sources—much more efficient and productive—will likely be found. One of those sources is sugarcane—a relatively close cousin of *Zea mays*. In Brazil, for example, most ethanol is produced from sugarcane, and more than half of the cars in the country burn it for fuel. It has now become a significant alternative to fossil fuels.[14]

The consumption of maize-stalk beer in Mexico declined after the 1600s, when sugarcane became a widely available source of sugar in the Americas, replacing many other plants—such as maize—as the base for alcohol production. The Tarahumara are among the few peoples in Mexico who continued to make cornstalk beer up to the twentieth century, when it had already been replaced in most other regions by sugarcane.[15]

Today, as in the past, the way we use materials is constantly changing, and the forces shaping those changes hinge both on individual decisions and preferences and on large-scale, global forces. We cannot assume that the current uses for today's crops and the means of producing them will resemble those in the future. Just as genetic research is helping us understand the process by which maize was domesticated, it will create new technologies—and even new maize—that will, perhaps, be far removed from the ways we understand the plant today. Two of the key questions that weave through this book are: How is our understanding of the past—seen through the lens of just one plant and the people who invented and used it—shaped by the present, and how is our view of the present itself shaped by this heritage?

In 2010 UNESCO inscribed traditional Mexican cuisine on its growing list of Intangible Cultural Heritage. Many indigenous Mexican foods, recipes, and cultural practices associated with this cuisine were recognized as being one of the world's heritage resources—honoring the farmers and cooks who gave the world some of its most important and sustaining foods. Foremost among these is maize. As UNESCO puts it on their website:

Traditional Mexican cuisine is a comprehensive cultural model comprising farming, ritual practices, age-old skills, culinary techniques and ancestral community customs and manners. It is made possible by collective participation in the entire traditional food chain: from planting and harvesting to cooking and eating. The basis of the system is founded on corn, beans and chili; unique farming methods such as milpas (rotating swidden fields of corn and other crops) and chinampas (man-made farming islets in lake areas); cooking processes such as nixtamalization (lime-hulling maize, which increases its nutritional value); and singular utensils including grinding stones and stone mortars. . . . Mexican cuisine is elaborate and symbol-laden, with everyday tortillas and tamales, both made of corn, forming an integral part of Day of the Dead offerings.[16]

Even in the most difficult of economic times, when the maize used to make tortillas and tamales may be difficult to afford, it will always hold a central place in the daily life of Mexicans and many other peoples in Latin America and around the world. Intangible heritage though it may be, maize is one of Mexico's tangible gifts from the past.

CHAPTER I

The Archaeology of Maize

THE DOMESTICATORS DOMESTICATED

The story of maize begins at least 9,000 years ago in southwestern Mexico as small groups of nomadic people found themselves attracted to stands of a rather tall, bushy tropical grass now known as teosinte (figure 1.1). We don't know what name these early indigenous Mexicans had for teosinte, but by the time of the Spanish Conquest there were many names for it, including *cincocopi, acecintle, atzitzintle.*[1] Today evidence of these first farmers and the teosinte plants they harvested is almost invisible—but we can see some traces left behind by the early descendants of both the plants and the people. For example, photographs of the tiny maize cobs, classified as *Zea mays* ssp. *mays,* that were found in Guilá Naquitz cave, Oaxaca, by Kent Flannery and his crew in the mid-1960s show parts of the earliest known individual plants that are descended from an ancestral teosinte plant (figures 1.2 and 1.3).[2] In order for these cobs, which are directly dated to 6,230 cal BP,[3] to have existed, not only did the ancient Oaxaqueños living near Guilá Naquitz cave have to have planted individual seeds, but their ancestors and neighbors also had to have planted and harvested teosinte seeds for hundreds of previous generations.

We do not know if these particular early Oaxacan maize plants themselves had descendants. After all, their seeds could have been completely consumed by people or animals and not gone on to propagate

FIGURE 1.1. Schematic drawing showing the shape of a modern hybrid maize plant (left) with two ears growing off the primary stalk, compared with a teosinte plant (right), which typically has many stalks or lateral branches and can have twenty or more small ears, or spikes. (Redrawn by Michael Blake after Beadle 1980:114. See also Lauter and Doebley 2002:335, figure 1.)

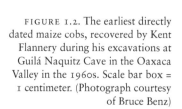

FIGURE 1.2. The earliest directly dated maize cobs, recovered by Kent Flannery during his excavations at Guilá Naquitz Cave in the Oaxaca Valley in the 1960s. Scale bar box = 1 centimeter. (Photograph courtesy of Bruce Benz)

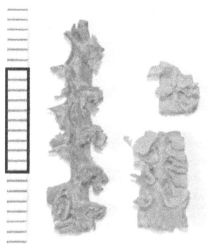

FIGURE 1.3. Never-before-published photo of excavations in progress at Guilá Naquitz Cave in the Oaxaca Valley in the 1960s. (Photograph courtesy of Kent Flannery and Joyce Marcus)

the next generation. Likewise, the people who planted and harvested these particular ears of maize and who carried them into the cave may not have gone on to have successful offspring whose descendants then gave rise to the present-day Zapotec and Mixtec peoples of Oaxaca. Even if neither the particular Guilá Naquitz maize plants nor the actual individuals who cultivated them contributed their genes to subsequent generations, we know that their cousins did. Maize and people still inhabit the valley, and they have even stronger reciprocal ties now than they did 6,000 or more years ago.

One of the key research questions about the origins of any domesticated species of plant or animal is who initially domesticated it? Related to this question is another: where did the domestication process take place, and, by extension, which subset of the wild population was domesticated? These are fundamental questions because their answers have the potential to release a cascade of knowledge about the general processes and specific mechanisms by which agriculture came about and spread around the globe. And, of course, these two questions are primarily about changing relations between humans and plants and animals.

The agricultural relationship between humans and other species is symbiotic in that both benefit from the process. Through this long-term symbiotic relationship, both the cultivators and the cultivated are "domesticated."[4] For example, it can be said that teosinte has benefited greatly by appealing to humans, who have spread it far from its natural homeland in Mexico to almost all regions of the world. This relationship is, however, a double-edged sword. In exchange for this "benefit," teosinte's domesticated descendant—maize—lost the ability to self-propagate, which means that individual maize plants are passively dependent on humans for their continued survival. The reverse is also true. We humans are dependent on maize (and hundreds of other domesticated plants and animals) for our survival. Every year millions of people around the world suffer death and disease from starvation and malnutrition because the plants and animals they and their ancestors have relied on are no longer available to them for various social, political, and environmental reasons (for example, because of warfare, repression, drought, or plagues). Deadly shortages can arise anywhere, and they can happen quickly.

By the same token, most of the human population of the earth is "domesticated" in the sense that we have been both socially and physiologically transformed by the domesticated plants and animals we rely on. To what extent is this process domestication? It is not domestication if we think of the process narrowly as the intentional manipulation of one species by another to select for characteristics or traits that are valued. Wild teosinte, wheat, rice, potatoes, apples, and a cornucopia of other plants were not intentionally raising humans to be better agricultural caretakers—producing farmers who knew about planting cycles, weeding, pest control, irrigation, and other useful agricultural practices. But then, to what extent were the very first teosinte harvesters trying to intentionally create corn on the cob or, for that matter, a global agribusiness centered on maize production? Neither is likely to have been the case.

Rather than stick to the standard definition of domestication as the adaptation of species to human environments, I will use the notion of reciprocal relationships between plants and humans, specifically maize and humans, and consider how those relationships varied through time and across landscapes, transforming people, plants, and landscapes in the process. These reciprocal relationships changed over generations and across space as they spread throughout the Americas. Maize domestication, like that of other plants, must have proceeded slowly and intentionally, but the intentions of the first maize farmers were likely quite different from those of later farmers and certainly different from participants in today's industrial agricultural systems. The first farmers' intentions with respect to maize had to be different because they were interacting with teosinte and early maize, which were structurally very different, as we shall see, from the varieties of maize produced by farmers several millennia later.

As many botanists have noted, the initial use of a plant, especially if that use involved selective harvesting and planting, automatically leads to changes—some intentional and others accidental—in the visible characteristics of future generations of that species. Planting and caring for the selected offspring of individual plants with preferred characteristics, such as larger seed size or less branching of the stems, leads to those same traits becoming increasingly dominant in following generations. This process inevitably leads to a transformation in the plant over time, whether or not there is any intentionality on the part of the domesticator.[5] Today we take this process for granted, as modern agriculturalists invest enormous resources in researching the most effective ways to improve crops so that their harvests will have the greatest chance of producing desired characteristics (increased yield, resistance to disease or drought, greater protein content, increased sweetness, and so forth). The first farmers, however, may have had somewhat different goals. They were mobile hunting, fishing, and gathering peoples, few of whom lived in permanently settled villages, yet they planted and harvested species of interest so that they would be available when and where they were needed. They were in all likelihood most interested in particular species of plants for their salient characteristics rather than for their unknown, and probably unimagined, potential future states. It is unlikely that the first teosinte and early maize farmers foresaw the need for, and tried to create, rot-resistant, high-yield, starchy, multicolored maize kernels with enhanced protein content capable of growing in latitudes far to the north and south of the plant's homeland. But if such characteristics appeared and were thought to be of interest, early farmers with exceedingly intimate knowledge of the life cycles of the species inhabiting their world certainly would not have ignored them and may have selected their seeds for future plantings.

One consequence of this process of interaction between humans and plants is that plants with the potential to generate variation can unintentionally and automatically influence the species that use them. Humans, or other animals drawn to early maize, could be transformed by these interactions by becoming habitual users. For example, a large concentration of any food resource has the potential to attract people who may eventually become dependent on it. Teosinte variants that more successfully attracted humans would themselves benefit from human interest by becoming increasingly prevalent in their environments—assuming that human interest resulted in replanting, weeding, watering, and other ways of nurturing the plant. Humans had to change their previous patterns of behavior to accommodate the changing plant, and in doing so they became reciprocally transformed in ways that they could not have predicted. It is in this sense that the domesticators become domesticated. We could think of domestication as biosocial entanglement—we become trapped in one another's webs of action and response, both behavioral and genetic.

Several botanists, anthropologists, and archaeologists have noted this reciprocal aspect of domestication. Usually though, when we talk about domestication we focus solely on the objects of human intervention and transformation, that is, the plants and animals that have been transformed. We typically discuss the transformation of people in terms of cultural evolution: the emergence of agriculture, and social, political, and economic complexity. So powerful is our image of ourselves as masters of our own history that it is difficult to imagine our utter dependence on the resources that we think we control. However, in light of the previous discussion, humans are resources for other species as well. From the standpoint of *Zea mays* ssp. *mays,* humans are agents of dispersal. If a sentient plant that wished to propagate itself and spread as far as possible could choose a species to manipulate, it could certainly do worse than choosing humans. By being genetically flexible, *Zea mays* has "persuaded" humans to move its seed around the globe faster and farther than any other plant in history. Maize's power over us is rather intimidating, and we cannot easily or practically release ourselves from its grip. In fact, maize is becoming, year by year, increasingly interwoven into our human existence. Our global human economy depends on it—just as *Zea mays* depends on us. Humans grow maize and maize grows humans.[6]

WHO DOMESTICATED TEOSINTE AND WHERE DID THEY DO IT?

Within the past few years, botanists have narrowed the search for the ancestor of maize and its natural range, and in the process they have indirectly pointed to the individuals who must have initially domesticated the plant. The "who" and "where" questions must first be preceded by "what": what plant was ancestral to the maize we know today?

Teosinte—Maize's Ancestor

For many decades the eminent botanist Paul C. Mangelsdorf and his students argued that modern maize arose from the domestication of a now-extinct wild maize, or pod corn.[7] Furthermore, he thought that maize might have had at least two separate origins, one in Mesoamerica and one in South America. This theory had long been in competition with another idea—championed by George Beadle and his students—that maize had arisen from the domestication of one of several subspecies of wild teosinte native to Mexico and Central America.[8] We will look more closely at this debate in chapter 3 because it opens the door to an amazing set of discoveries that demonstrate the interdisciplinary nature of our quest for knowledge about agricultural origins. But for now, the short version of the story is that the extinct wild maize hypothesis has been rejected, and the teosinte hypothesis has been confirmed by many new and independent lines of evidence.

Until recently it was not certain whether one or both of the two main annual subspecies of teosinte (Chalco and Balsas populations) were the ancestors of maize. But now, thanks to the explosion of new genetic studies, this has been mostly resolved. It is now widely agreed upon that all maize is primarily descended from one subspecies of annual teosinte—*Zea mays* ssp. *parviglumis*—found most commonly in the Balsas River region of southwestern Mexico—hence its original name, Balsas teosinte.[9] But plant geneticists have discovered that Chalco teosinte—*Zea mays* ssp. *mexicana*—has also contributed genetically to modern maize, which, as a result, contains genetic traits of both.[10] These two subspecies are very closely related and in fact live as **sympatric** neighbors—with *parviglumis* inhabiting lower elevation terrain and *mexicana* living at higher elevations (map 1.1).

How do these two subspecies of teosinte fit into the overall genus *Zea?* Botanists, relying on the morphological characteristics of teosinte varieties in Mexico and Central America, have defined two main groups, or "sections," of the genus *Zea* (figure 1.4). One group is called Section Zea and includes one species, *Z. mays,* and all of its four subspecies: *mays* (modern domesticated maize) and *parviglumis, mexicana,* and *huehuetenangensis* (three annual teosintes). The second group is called Section Luxuriantes and includes four species: *Z. luxurians, Z. nicaraguensis, Z. diploperennis,* and *Z. perennis.*[11] All but the last two members of *Zea* are **annuals**—both *diploperennis* and *perennis* are **perennials,** as their specific Latin names readily suggest. The morphological

MAP 1.1. The present-day distribution of six species and subspecies of teosinte in the genus *Zea*. Modern maize is descended primarily from *Zea mays* ssp. *parviglumis,* with introgression from *Zea mays* ssp. *mexicana*. (By Michael Blake and Nick Waber, adapted from Fukunaga et al. 2005:2242, figure 1)

traits that prompted this classification scheme have been backed up with genetic analyses that show the interconnections among the members of the *Zea* genus yet track a long genetic history that extends back tens of thousands of years. In spite of this long history, the botanist John Doebley and his colleagues point out that, except for *parviglumis* and *mexicana,* the species and subspecies of *Zea* have relatively limited geographic distributions. *Parviglumis* and *mexicana* have both greater distributions and greater morphological variation, and, although we need not elaborate on this here, it is worth mentioning that they each have a number of "races" or varieties restricted to specific parts of their geographic range.[12]

Parviglumis thrives in seasonally moist habitats between about 400 and 1,800 meters above sea level, and, although it is most common in the central and eastern Balsas River region in the states of Michoacán,

FIGURE 1.4. The phylogenetic relationships of six species and subspecies of teosinte in the genus *Zea* and their relation to a close cousin, *Tripsacum*. Maize (*Z. mays* ssp. *mays* not shown) is most closely related to *Zea mays* ssp. *parviglumis, mexicana,* and *huehuetenangensis* in Section Zea. *Diploperennis, perennis,* and *luxurians* are grouped in Section Luxuriantes (*Z. nicaraguensis* not shown), and *Tripsacum* is a separate, but related genus. (Drawing by Michael Blake in Hart et al. 2011:2, figure 1)

Guerrero, and México, it also occurs in the neighboring states of Jalisco and Colima to the northwest and Oaxaca to the south.[13] Its distribution may have been greater in the past under different environmental conditions, but not enough evidence exists yet to test this hypothesis.

Mexicana, the other main subspecies of teosinte, grows to the north and west of *parviglumis*'s home range, in the states of Michoacán, Guanajuato, Jalisco, and México. *Mexicana* is adapted to higher elevations (between about 1,600 and 2,700 meters) and drier conditions than *parviglumis.* During periods of increased aridity in the past, *mexicana* may have spread to lower elevations, displacing *parviglumis,* which prefers more rainfall.

Parviglumis (which means small-**glumed**) was scientifically described and classified by botanists Hugh Iltis and John Doebley more than thirty years ago.[14] The samples they described in detail grow in wild stands on the south-facing slopes of the Balsas River region, some 220 kilometers west of Mexico City. Iltis recounts the story of *parviglumis*'s "rediscovery" in several publications, but perhaps the most humorous telling was at a Society for American Archaeology symposium held in Montreal in 2004:

> My fellow mutation hunter Ted S. Cochrane and I, in December of 1971, in this ancient cradle of maize domestication, south of Morelia on the high slopes of the Meseta Central escarpment overlooking the Rio Balsas valley and just south of the little pueblo of Tzitzio, (and on a hint from a young,

black-haired, sharp-eyed, and lively native American chambermaid at Motel Morelia in the city of that name while watching us make herbarium specimens of teosinte) we discovered several fine stands of our special grass, both truly wild ones on almost vertical rocky slopes and weedy ones in maize fields, where, with tassels and leaves already removed by the local *campesinos* to feed their cattle, the dried out maize ears were ready for harvest. It may well have been nearby that some eight millennia ago, perhaps even in that same teosinte patch, one then already well-known and becared by the keen mentality of an unsung, ever-hungry people, this naked-grained mutant made its unexpected appearance, one that surely would not have gone unnoticed for long. In fact, we may imagine that maize domestication may well have begun here with the startled cry in Nahuatl of some bright, strong, young Indian woman or man, a "Xilonen" or a "Cuauhtemoc," holding a cluster of young, crisp mutated teosinte ears in hand, exclaiming excitedly to a companion, "Look, look what I found—these surely must be *teo centli!*"[15]

Balsas Teosinte and Its Early Cultivators

So far, the weight of evidence suggests that *Zea mays* ssp. *parviglumis* was the first ancestor of all modern maize and that it was domesticated by the people of the Balsas River region, beginning at least 9,000 years ago. *Parviglumis*'s role as the initial progenitor of all domestic maize was only recently established by Yoshihiro Matsuoka, John Doebley, and their colleagues in a remarkable paper published in the *Proceedings of the National Academy of Sciences* in 2002.[16] Matsuoka, the lead author of the study, was, at the time, a postdoctoral fellow in the famous Doebley Lab in the Department of Genetics at the University of Wisconsin-Madison, where many of the major discoveries about the genetic ancestry of maize have taken place, including the definitive studies showing that teosinte was the wild ancestor of modern maize. In the 2002 study the team looked at the genetic similarities and differences among an enormous sample of different races of maize and teosinte collected from plant populations living in North and South America and found that all modern maize was genetically most similar to *Zea mays* ssp. *parviglumis* and more distantly related to other subspecies of teosinte. We will look more closely at this study and other new discoveries about maize's genome in chapter 8.

Genetic studies of maize and teosinte carried out during the past decade are showing, as geneticists and botanists such as George Beadle and Hugh Iltis had long argued, that the first maize farmers must actually have been teosinte farmers. Furthermore, unless the range of *parviglumis* has changed radically during the past ten millennia, it is likely that the first *Zea mays* domesticators were the aboriginal occupants of the Balsas

River region. The archaeology of this region is not well known, and, compared with the Tehuacán Valley cave sites where Scotty MacNeish excavated, or those in the Valley of Oaxaca where Kent Flannery's discoveries were made, very few detailed excavations have been carried out. I expect that some day archaeological research will turn up early teosinte use in the Balsas River region. But so far, except for a few rare examples, there is little archaeological evidence of the first teosinte farmers actually using, processing, and discarding teosinte remains in any ancient sites.

Guilá Naquitz Cave in the Oaxaca Valley still holds the record as the location where the oldest distinctly recognizable, unequivocally dated maize cobs were discovered.[17] The cave lies at about 1,925 meters in elevation—just above the altitude where the natural environmental range of *parviglumis* and *mexicana* overlap. Guilá Naquitz Cave was occupied as early as about 10,000 years ago, but the maize cobs are much more recent—about 6,230 years old, give or take a century. These tiny, ancient cobs are a good example of what archaeobotanists call "macroremains," pieces of ancient plants that are large enough to be observed by the unaided eye and that, if complete and well-enough preserved, can often be identified to species and variety. Still, as old as these cobs are, it is unlikely that they represent the earliest domesticated teosinte. Even though they share many characteristics with teosinte, and so represent a relatively early stage in the evolution of modern maize, genetic studies strongly suggest that there were at least three thousand years of experimentation with maize farming in order to get from wild teosinte to domesticated early Guilá Naquitz maize. We will return to this question in much more detail in chapter 5.

Most botanists who study the origins of maize are now fairly certain that *parviglumis* was first domesticated farther north in the Balsas River region of Guerrero and Michoacán rather than in Oaxaca. This is because the Balsas region, and west-central Mexico in general, have the greatest degree of genetic diversity of teosinte. For decades botanists have thought that geographic regions where we see a plant's highest genetic diversity are most likely to be the locations of that plant's initial domestication. In recent years this has drawn researchers to look for evidence of early cultivation in the Balsas region. A previous generation of botanists and archaeologists who thought that maize must have originated from *mexicana*, or a putative wild maize species, looked for the origins of maize in higher-elevation locations on Mexico's central plateau.

It now appears that, based on several lines of evidence, including those produced by new genetic research, both subspecies of teosinte—

parviglumis and *mexicana*—contributed significantly to the genetic diversity of modern maize. It may be that after the initial domestication and spread of *parviglumis,* early "parviglumoid" maize interbred (through a process known as **introgression**) with *mexicana,* creating hybrids with genetic traits that were somewhat different from either subspecies of wild teosinte and distinct from modern maize. Some of these traits may have helped *parviglumis*-cum-early maize adapt to higher elevations and more arid conditions as the earliest farmers moved their new creation outside of its homeland. One recent study has shown that 2–4 percent of maize's genetic variants came from *mexicana,* while even newer research suggests the contribution may be somewhat higher.[18]

Who were these first long-term teosinte users who lived in the Balsas River region of Guerrero, Jalisco, and Michoacán? What do we know of their history and archaeology? The short answer is, unfortunately, very little.[19] The earliest period that is well dated is called the El Opeño phase, beginning around 3,500 years ago and represented by spectacular shaft tombs and their amazing ceramics and other grave offerings. The long time span before the El Opeño phase is usually called the Archaic period, which is typified throughout Mesoamerica by the rather sparse remains of hunting, fishing, gathering, and farming peoples who, for the most part, did not yet live in permanent villages. In the uplands, above 400 meters in elevation, there are very few archaeological sites with evidence of Archaic period occupants. For example, in a recent search for Archaic period agricultural sites in Jalisco, my colleague Bruce Benz—a botanist who trained under Hugh Iltis and also happens to be an archaeologist—visited thirty-six rockshelters and fifteen open-air sites in the Sayula-Zacoalco Basin just to the southwest of Lake Chapala (almost 500 kilometers west of Mexico City).[20] Benz undertook excavations at three rockshelters that he thought might have undisturbed deposits. In one of them, Abrigo Moreno 5, he was able to radiocarbon date charcoal from one of the lowermost layers and found that it was about 5,500 years old. Unfortunately, however, he did not find any remains of maize associated with this deposit—perhaps if maize or other plants had been used there they had long since decayed.

HOW AND WHEN DID EARLY MAIZE SPREAD?

At some point between about 9,000 and 6,200 years ago, during the Archaic period and in the general vicinity of the Balsas River region, a mutant form of teosinte showing the rudimentary signs of maize must have appeared. Judging by the size and characteristics of the Guilá Naquitz cobs, the first ears of this mutant teosinte must have been tiny, but they must also have had a propensity for solid, non-shattering cobs. These earliest versions of maize also may have had ears that were enclosed within a leafy husk.[21] This would have resulted in the plant losing its ability to self-propagate, because the seeds would not have been able to separate from the cob when ripe and would not have been able to get free of the husk. This would mean that the first teosinte-maize growers would have had to have sown and tended each season's crop. Future generations of teosinte-maize seeds with these traits could only be viable with direct human intervention in the growing cycle. The Guilá Naquitz maize shows that Archaic peoples had already been intervening in the seed selection and planting process for many hundreds of generations before it arrived at the cave.

Early Archaic period peoples who first used and domesticated teosinte probably traveled a great deal during the course of each year, but even so, they must have regularly planted and tended a range plants in their seasonal cycles that were of interest to them. Early maize was just one of several species that they planted and harvested, and it had to find a niche within the broader system of plant harvesting. Other plants that they used that later became important domesticates included squash (*Cucurbita* sp.) and beans (*Phaseolus* sp.).

How did early maize spread from group to group, eventually moving beyond its natural range? The two main possibilities that come to mind are: (1) people with a knowledge of how to cultivate maize spread outward from teosinte's homeland, taking early maize (in other words, recently domesticated teosinte) with them, and (2) people traded or gave maize seeds to their neighbors, who in turn passed them on to their neighbors, and so on—a form of "down-the-line" exchange. The first scenario makes sense if early maize provided a nutritional advantage to people that allowed their populations to grow more rapidly than their neighbors or allowed them to move into previously unoccupied regions. This form of farming expansion is referred to as **demic**—an expansion of population into neighboring territories.[22] In the first model, early maize would have been a new and significant addition to subsistence economies, while in the second model it would have been an interesting and attractive but nonessential supplement. The second scenario is reasonable to expect if early maize seeds were just one more thing that was traded among peoples who ranged over large regions and who were already well settled in a diverse set of environments.

I think that, while these two scenarios are not mutually exclusive, it is unlikely that the first model describes what happened during the initial centuries or few thousand years of maize's domestication and use. It may well describe how more-developed maize farming, and farmers, spread in many regions of the Americas long after maize had been domesticated and after it had become much more similar to the highly productive grain crop we know today. During maize's initial period of domestication and spread it is much more likely that maize moved in a down-the-line fashion, being traded or gifted among hunting, fishing, and gathering peoples who were also part-time horticulturalists, tending a range of plants that were important to them for a variety of reasons (including their uses as food, technologies, and medicines).

There is considerable archaeological evidence that Archaic period peoples of Mexico had well-developed interaction networks and that goods moved over long distances. The best examples of goods traded through these networks are objects made of stone and marine shell. Because they can be preserved indefinitely, stone tools leave the clearest fingerprint of this exchange. Obsidian is the primary stone type used to illustrates these long-distance interactions because it can be so precisely linked to a few well-known source locations—places where this glassy, super-sharp, igneous rock can be quarried. But it is not the only stone. Chert, a fine-grained sedimentary rock prized for its durability and the ease with which it can be shaped by chipping, is also identifiable to source location, and there is evidence of this stone type having been traded over long distances as well. Besides stone tools, marine shells crafted into valuables, such as beads, bracelets and pendants, made their way from the coast to the interior. Although much rarer than stone artifacts, these shell objects are useful in documenting early contacts between coastal lowlands and interior regions—contacts that must have been in place if early maize was spread by way of exchange networks.

The preservability of stone and shell contrasts with the perishable nature of most botanical materials. Large, readily identifiable pieces of plants do not usually preserve in exposed archaeological sites unless they are charred. Outside of dry cave sites, no large pieces of early maize (charred or otherwise) have been reported for Archaic period archaeological sites in Mexico and Central America. But as we will see in chapter 6, microscopically small plant remains do preserve in the form of pollen grains, phytolith particles, and starch grains. These all have distinct sizes and shapes that can often be identified to the species level and provide a reliable way of determining the presence of plant use in the

FIGURE 1.5. The Xihuatoxtla Rockshelter, located in Guerrero, Mexico, which has produced maize microremains (phytoliths and starch grains) dated to as early as 8,750 years ago. Located in the heartland of maize's wild teosinte ancestors, this site was excavated by Anthony Ranere, Dolores Piperno, and their team, producing physical evidence that teosinte was domesticated around the same time that had been suggested by genetic research. (Photograph courtesy of Anthony Ranere)

absence of plant macroremains. At the site of Xihuatoxtla, a rock shelter located in the upper Balsas River region of Guerrero, Dolores Piperno, Anthony Ranere, and their colleagues report finding maize starch grains and phytoliths that are associated with charcoal dated to 8,750 years ago (figure 1.5).[23] Extending southward and eastward from the Balsas region, researchers have discovered maize microremains in a dozen or so locations—some associated with archaeological sites but most occurring in natural deposits, such as lake and swamp sediment cores. Some of these remains date indirectly to between 8,300 and 4,500 years old and have been found in sites ranging from Mexico and Central America to northern South America, most often in regions within a few hundred meters in elevation above sea level.[24]

These discoveries of maize microremains outside of the natural range of the teosintes that were maize's ancestors are remarkable in many ways. First, they suggest that mobile Archaic peoples readily accepted teosinte-like maize very soon after it was first domesticated in the Balsas region. Second, they confirm that this very early (more than 6,000-year-old) maize was still teosinte-like because it could not yet have had time to transform—through agricultural selection—into the larger-cob maize that we know so well from much later time periods. Third, archaeological sites with pre-6,000-year-old maize microremains occur most commonly in the coastal lowland regions of Mexico, Central America, and South America. We do not yet have much evidence of very early maize microremains from sites in the highlands—possibly because these types of microremains do not preserve as well in such environments, but their relative absence is more likely a result of the fact that this research is still in its infancy.

Also in its infancy is the direct dating of maize macroremains (fragments of cobs, kernels, stalks, and leaves) from early time periods. It has been little more than twenty-five years since the first of these remains were directly dated using accelerator mass spectrometry (AMS) radiocarbon dating, a technique that allows researchers to use very tiny samples of an ancient plant—for example, part of an individual kernel—to determine its age.[25] Prior to the development of this method, archaeologists were forced to date whole cobs (thereby destroying them and foreclosing any possibility of further study) or to date materials such as charred wood from the same layer or deposit and assume that the maize found in association with the dated charcoal was roughly the same age. This assumption has proven dubious in many cases, as we will see in chapter 4. So far we do not have any actual remains of teosinte-like early maize outside of the dry caves of Mexico. In fact, the maize from both the Tehuacán Valley caves and the Tamaulipas caves is as much as 1,500 years younger than the Guilá Naquitz maize from Oaxaca and much younger still than maize pollen, phytoliths, and starch grains reported from as far away as Panama and Ecuador.

One implication of the pattern of very early movement of teosinte-like maize into regions of the Americas, far distant from teosinte's natural range, is that it must have been of great interest to nomadic or seminomadic peoples who made their living by hunting, fishing, plant gathering, and some cultivation, and who must have been in contact with one another (even if indirectly) through vast networks of exchange relationships. Another, and perhaps even more important, implication is that these first importers and growers of early maize must not have been interested in the plant for its high-yield, large ears full of grain because maize did not yet have such features. It is possible that very early maize had many small ears per plant—teosinte can have between ten and one hundred small ears—and, if so, this might have been what attracted early farmers. But this seems unlikely because one of the first mutations that took place in domesticated maize was the appearance of *teosinte branched1*, the gene that suppressed teosinte's lateral branching from the main stalk and led, in maize, to the condensation of the branches into a **polystichous** form with only one or two ears nestled tightly against the main stalk.

This still leaves us facing a major puzzle: what was it about the early teosinte-like maize plant that attracted so much interest during the period between about 9,000 and 4,500 years ago, when the plant spread so far afield, even though its ears were still so tiny? Could it be that people were more interested in the green immature ears, which were both sweet and nutritious? Or were they keen to use the stalks of the plant, from which they extracted a sugary juice? Perhaps they plucked the ears off to eat fresh, allowing the sugars to accumulate in the stalks, which could then be squeezed to extract the juice, similar to what was done with sugarcane.[26] This juice could then be fermented for a few days to produce an alcoholic beverage—just as the Rarámuri people of northwestern Mexico did with maize stalks until recent times.[27] Regardless of the initial reasons for the early Mesoamericans' interest in using and spreading the first domesticated teosinte, the plant morphed over the course of its first several millennia of interaction with humans so that people eventually became much more interested in its grain-bearing ear than what the stalks had to offer.

HOW DID MAIZE TRANSFORM UNDER CULTIVATION?

What do we know about the transformation of maize after about 6,200 years ago—the age of the earliest securely dated maize cobs recovered and described so far? Maize, because it is so genetically flexible, was modified by peoples who lived in many different environments and had many different cultural preferences and practices. If the initial spread of maize was by exchange (diffusion) rather than by population (demic) expansion, then its malleability leant itself to manipulation by these different peoples. At least in one region it appears that early farmers selected maize for its cob size and grain yield. This we know from the Tehuacán Valley sequence and the new work of Bruce Benz and his students, which built on the earlier discoveries of Paul Mangelsdorf in the 1960s and 1970s.[28] In chapter 5 we will look at studies of maize macroremains from these dry cave sites and examine the evidence for the impact of selection on the transformation of the ear.

As fascinating and revealing as such studies are, however, they do not tell the whole story. Aside from the durable cob, not many parts of the plant are well represented in the archaeological record, and even where they are—such as, for example, at some of the Tehuacán Valley caves—the stalks, leaves, roots, and husks have not been as extensively studied as the cobs have.[29] Eventually these parts of the plant must also be studied in detail so that we can see how other physical (phenotypical) characteristics of maize were modified under cultivation. Archaeological sites along the arid coast of Peru have yielded early maize remains that provide such details. There are dozens of sites spanning the period from about 3000 BP to the Spanish Conquest that have maize remains, including kernels, cobs, and various other plant parts. Remarkably, all the samples recovered so far show a similar range of evolutionary changes to those observed in Mexico. Later period Peruvian sites dating to the fourteenth century show maize that is well developed and very much like the late period maize found from eastern North America to the US Southwest, through Mexico, down to Central America and throughout South America.

This pattern suggests that, although maize evolved within each region and continued to evolve as it was moved farther and farther from its homeland, its primary characteristics (one main stem with a small number of ears, non-shattering cob, and naked kernels) were already long fixed. Eventually other characteristics, such as kernel quantity, shape, size, and color, were carefully nurtured by farmers with different cultural preferences in different regions throughout the Americas. Characteristics that weren't visible, such as sugar content, starch quality, protein type, resistance to rot and pests, response to day length during the growing season, and so on, were also carefully selected. These characteristics are much more difficult to determine from archaeological macroremains and in many cases must be inferred from variations in present-day collections from a range of geographic locations. Genetic analysis of these variations has been pivotal in determining the relationship among varieties of maize and establishing that they all, regardless of their outward appearance and less salient traits, have a single common ancestor.

The unconscious process of domestication in seed plants is generally referred to as the "adaptive syndrome of domestication," and has been observed in many species of agriculturally important plants. This syndrome is a constellation of trait changes that come about as a result of intentional actions on the part of the humans who propagate, tend, and harvest the plants they are interested in. Bruce Smith, an archaeologist at the Smithsonian Institution who specializes in archaeobotany, has recently described this process. He summarizes the five major changes that are expected during domestication as follows:

(1) simultaneous ripening of seeds;
(2) compaction of seeds in highly visible terminal stalk/branch "packages";

FIGURE 1.6. Multicolored, dried ears of maize—the best of the year's crop saved for the next season's planting—hanging in the rafters of a house in the Tzeltal Maya town of Chanal, Chiapas, Mexico. (Photograph by Michael Blake)

(3) seed retention (loss of natural seed dispersal mechanisms);
(4) increase in seed size; and
(5) simultaneous and rapid seed germination (loss of germination dormancy, reduction in seed-coat thickness).[30]

Most of these changes can be considered the unintentional consequences (albeit desirable from the farmer's point of view) of intentional interventions in the lifecycle of target populations of plants. As Smith so clearly describes, it is the storing of seeds during part of the year for planting under carefully controlled conditions at a future date that creates the environment for new and nonnatural selective pressures (figure 1.6). These human-induced selective pressures, whether applied to maize, beans, or squash, lead to the same sorts of archaeologically visible changes. This "adaptive syndrome" would have the same constellation of impacts even if the farmers were ultimately interested in consuming or using parts of the plant other than the seeds. We would expect this domestication syndrome for maize, gourds, cotton, and any other plant where the means of storing, planting, and harvesting the seeds were crucial to the cultivation process. Therefore it is not, strictly speaking, necessary for the first maize farmers to have been consciously selecting for increased grain size or cob size for the cobs and grains to increase in productivity.

By about 3,000 years ago maize in the southwestern United States, and from northern Mexico to Central America and some regions in South America, had become a dietary staple whose dry and stored grain was likely more important for food than any other uses the plant may have had. At this point maize was a larger-cobbed plant than it had been during the previous millennia, and it had had many hundreds or thousands of generations to adapt to each region in which it had been introduced. For example, in Arizona and New Mexico, some 2,000 kilometers to the north of teosinte's homeland, maize had been around for at least a thousand years by this time and was becoming an increasingly productive and important part of peoples' systems of food production.[31] In central Mexico, maize had increased in size and was clearly on its way to becoming ever more productive.[32] In southern Mexico, humans had started to show the impact of maize in their diet—it became increasingly important and left a characteristic chemical signature in people's bones and teeth (a topic we will examine more closely in chapter 7). About this same time, and some 4,000 kilometers farther south along the Pacific slopes of the Andes, maize was being used both as a food and, even more importantly, for making maize beer—*chicha*—and as such had a growing dietary and ritual importance.[33] Its significance in the making of *chicha* began at least by the Early Horizon period, about 2500 BP, and continues to the present day.[34] Whether for food or drink, maize took on important spiritual meanings and ceremonial significance—eventually becoming intimately associated with the social identity of most of the Native American peoples who grew it.[35] For many, this once humble grass from western-central Mexico belonged, and for some still belongs, to the spiritual realm—linking living people to both their ancestors and their gods.

CHAPTER 2

The Place of Maize in
(Agri)cultural Origin Stories

In view of the importance of the agricultural transformation
of the earth, it is not surprising that scholars have long been
interested in the origins of agriculture.

—Bruce D. Smith

Agriculture plays an increasingly important role in Western origin stories, which show how we see ourselves and give explanations for how
we got to be the way we are. The stories of how present-day societies
came to be are very often intertwined with the tales of how agriculture
originated. These accounts are seldom neutral. The practice of agriculture is perceived as being much more than the simple task of providing
food for our daily sustenance—more often than not, it frames and
defines society's moral values and fundamental principles. In Western
societies our cultural history places a strong emphasis on particular
kinds of food and methods of production as being significant, appropriate, and valued. These are generally the foods and production methods
that, thousands of years ago, originated in and were dispersed from the
Near East—especially the Fertile Crescent, stretching from the Mediter-
ranean coast in the west to the Tigris and Euphrates Rivers in the east,
long considered the cradle of Western civilization. Seen through this historical and cultural lens, most other kinds of food production, whether
agricultural or nonagricultural, were, and still are, often considered
marginal, if not outright immoral.[1] Therefore, it is not surprising that
these modes of thinking have colored the way that archaeological
research about the origins of agriculture has proceeded. Archaeologists
also have cultural biases, including specific food interests. In particular,
these biases have led a generation of researchers to emphasize models of
agricultural origin that are appropriate for the Near East. Until recently,

there was a hesitancy to examine the histories of foods and food-production systems that are characteristic of other regions of the world and other time periods.

Keeping Western perspectives in mind, including Western archaeologists' views of agricultural origins as human origin stories, allows us to think about the relevance of origin stories in general—and food origin stories in particular—as a way of understanding the variety and complexity of New World food crop histories. In this chapter, we will explore the role agriculture origin stories have played in the development of archaeological thought, analyze how these stories have impacted research on the origins of agriculture in the New World (specifically how they have shaped the main theories of maize agriculture), and look at some examples of indigenous origin stories that link to agriculture in general and maize agriculture in particular.

AGRICULTURE AND CULTURAL EVOLUTION

Questions about how agriculture began and why it became so important to many human societies have been central concerns of anthropology and archaeology for a century and a half. In American archaeology this interest in—one could almost say preoccupation with—agriculture stems from anthropology's early roots in cultural evolutionary theory in the nineteenth century. But looking back even further, the influence of Western European social theorists in the 1600s and 1700s laid the intellectual foundations for thinking about the connections between modes of subsistence and modes of social, political, and economic organization. Many excellent works survey this history of anthropological thought and show the fluctuating strength of the links between subsistence and social categories and the impacts of these linkages on academic and popular thought.[2]

In the evolutionary scheme, agriculture is given pride of place because it is intimately linked with the way we conceive of our society's very existence. It is impossible to imagine most present-day societies—urban or rural—being able to survive for more than a few weeks or months without the benefits of food production. There are very few people today who supply all their own food and other material needs without relying on agriculture. Up to the end of the last ice age, approximately 10,000 to 12,000 years ago, the reverse was true—most people depended on hunting, fishing, and gathering for their daily subsistence needs. Even though this seems like a long time span to us now, it is really rather short in the overall continuum of human history, especially if we consider that fully modern humans have been around for one hundred thousand years or more.[3] This means that explanations of the appearance of agriculture have to take into account the transformations that took place during the past ten thousand or so years and scrutinize the new ways that people interacted with the plants and animals in their environments.

The emergence of agriculture is a topic that continues to fascinate researchers because of the enormous variety of ways that peoples living in diverse environments chose to interact with so many different species of plants and animals. Some of these interactions led to agricultural production systems and others did not. We have come to see this as one of the great divides in human history. One of the outcomes of the divergence between emerging agriculturalists on the one hand and continuing hunter, fisher, gatherers on the other hand is that the former developed ever increasing and expanding populations—eventually absorbing, displacing, fighting with, or killing off (directly or indirectly) non-food-producing peoples. These farmers tended to think that their way of existing was superior to that of the societies they replaced, and they linked their economies—including agricultural production systems—to notions of morality, energy, progress, divinity, and the very definition of civil society. Indeed, agriculture is one of the defining elements of civilization in both scholarly and popular literature.

The flip side of "civilized" is "uncivilized." From the point of view of farming-dependent, often urban-dwelling, folk—some of whose descendants developed the very discipline of anthropology—"uncivilized" meant less developed. In fact, beginning in the 1700s the term became synonymous with "savage" and "barbarous." In English the "wildness" denoted by the adoption of the French term *sauvage* perfectly captures the classification that early social theorists intended. One could argue that the definition of civilization as a category relies on a contrast with wildness— untamed people with "rude" customs and practices, living in "primitive" social arrangements and representing a "primal state" of human existence. One of the central aspects of Western imaginings of this savage existence is the description of what were seen to be rudimentary forms of economy, such as subsistence practices.[4] Discussions of such peoples could be reduced to the following formulation: wild people exist on wild plants and wild animals because neither the people nor the plants and animals that they relied on were domesticated. They, and the very landscape they inhabited, were seen to be uncultivated, untamed, and undeveloped—in a word, "wild."[5]

These perspectives have been laid out and elaborated in dozens of books, beginning with one of the very first anthropology textbooks, published in 1865 by E. B. Tylor, an early British social anthropologist.[6] Most texts produced within the past decade or two continue to reiterate these views, albeit in a more modern form. Bruce Smith writes on the topic in his superb synthesis *The Emergence of Agriculture:*[7] "When these new agricultural economies emerged, they didn't just allow human population growth, they also fuelled the creation of ever larger and more complex human societies, far beyond what had developed in hunting and gathering times. . . . Villages turned into towns, cities and city states gained control of growing agricultural landscapes, and empires emerged as our ancestors became more and more successful at organizing agricultural production and the populations it fed."

Smith's view of the significance of agricultural production is based on decades of meticulous research—data collection, analysis, and interpretation—which has created a rich repository of information about the history of food production systems that was simply not available to early theorists. Yet the general sweep of human "progress" implied in this passage is not unlike that implied in dozens of earlier depictions, which were not based on much, if any, actual data. As Mark Pluciennik points out in his analysis of the history of subsistence categories in the definition of hunting and gathering societies, earlier American anthropologists held similar viewpoints.[8] One of these, Alfred Kroeber, wrote in the 1930s, "Much more important than the ground stone axe in its influence on life was the commencement, during the Neolithic, of two of the great fundamentals of our own civilization: agriculture and domestic animals. These freed men from the buffetings of nature; made possible permanent habitation, the accumulation of food and wealth, and a heavier growth of population."[9]

Earlier social historians and philosophers, such as Karl Marx and Frederick Engels, contrasted societies that did not practice agriculture with those that did as a central distinction in their discussion of the nature of political economy.[10] They relied heavily on the anthropological research of Lewis Henry Morgan, one of the founders of American anthropology in the mid-1800s. Morgan defined three stages of human progress—savagery, barbarism, and civilization—based on technological and material systems as well as social organization. In creating his scheme, he called agriculture one of the "arts of subsistence," using it to define the notion of "advancement" up the evolutionary ladder: "The cultivation of cereals and plants was unknown in the Western hemisphere except among the tribes who had emerged from savagery; and it

FIGURE 2.1. Pieter Bruegel the Elder, *The Harvesters,* 1565. This painting exemplifies the landscape and labor of grain harvesting in Western Europe in the mid-1500s and is contemporaneous with the initial European conquest of indigenous peoples in the Americas. (The Metropolitan Museum of Art, Rogers Fund, 1919. OASC: www .metmuseum.org/collection/the-collection-online.)

seems to have been unknown in the Eastern hemisphere until after the tribes of Asia and Europe had passed through the Lower, and had drawn near to the close of the Middle Status of barbarism."[11]

The pinnacle of cultural achievement for these early social theorists was, of course, civilization—and for Morgan and other social evolutionists of his day, agriculture, especially the cultivation of cereals and the domestication of animals, was an important foundation of the Western-centric view of progress and civilization (figure 2.1).

A century before Morgan, the Scottish social philosopher Adam Smith (figure 2.2) used the same terms to describe indigenous peoples in the Americas in his seminal 1776 work, *An Inquiry into the Nature and Causes of the Wealth of Nations:*[12]

> New Granada, the Yucatan, Paraguay, and the Brazils were, before discovered by the Europeans, inhabited by savage nations, who had neither arts nor agriculture. A considerable degree of both has now been introduced into all of them. Even Mexico and Peru, though they cannot be considered as

FIGURE 2.2. Portrait of Adam Smith (1723–1790). Smith was widely acclaimed for his book *The Wealth of Nations*, the founding text of the nascent field of political economy, in which he commented frequently on "Indian corn" farming in the Americas and the supposed inferiority of New World agriculture. (Photograph of an engraving by John Kay, courtesy of the Library of Congress, www.loc.gov/pictures/item/91706325)

altogether new markets, are certainly much more extensive ones than they ever were before. After all the wonderful tales which have been published concerning the splendid state of those countries in ancient times, whoever reads, with any degree of sober judgment, the history of their first discovery and conquest, will evidently discern that, in arts, agriculture, and commerce, their inhabitants were much more ignorant than the Tartars of the Ukraine are at present.

The Native Americans Smith mentioned, all of whom practiced their own forms of agriculture, were, in his estimation, not really agriculturalists. This is, no doubt, because they didn't farm in the European fashion—that is, they didn't grow wheat or raise cattle. It is easy to see how this framework was devised within the cultural milieu of European and North American industrializing agricultural societies that, in the eighteenth and nineteenth centuries, were still in the process of trying to absorb, assimilate, or annihilate the "savage" and "barbaric" peoples of the Americas and other regions of the world. Smith makes frequent mention of agricultural progress and expansion in the American colonies—and he seems impressed with the economic potential these new "unoccupied" lands have for colonial agricultural expansion: "It has been the principal cause of the rapid progress of our American colonies towards wealth and greatness that almost their whole capitals have hitherto been employed in agriculture."[13] Smith goes on to observe: "The colonists carry out with them a knowledge of agriculture and of other useful arts superior to what can grow up of its own accord in the course of many centuries among savage and barbarous nations."[14]

The expansion of European colonists into the western "frontier"— the very notion of which implies a deep concern with the moral duty to tame landscapes that were perceived to have been underused or not used at all—essentially defines the post-European contact history of North and South America. The significance of indigenous food production— whether it involved agriculture or hunting, fishing, and gathering—was very often downplayed and undervalued. Even in the last half of the eighteenth century, crops such as Indian corn and potatoes were thought by most Europeans to be inferior foods. The following description by Adam Smith captured the European perception of food production in the West Indies: "The vegetable food of the inhabitants . . . consisted in Indian corn, yams, potatoes, bananas, &c. plants which were then altogether unknown in Europe, and which have never since been very much esteemed in it, or supposed to yield a sustenance equal to what is drawn from the common sorts of grain and pulse, which have been cultivated in this part of the world time out of mind."[15]

For Europeans, who began the process of colonizing the Americas in the late fifteenth century, grains such as wheat and barley—along with a commitment to animal husbandry—always counted the most.[16] From the outset, colonial encounters with indigenous Americans provided the colonizers with opportunities to observe local foods and food production practices, and, more often than not, they found them wanting, if not downright immoral. The settlers clung to the idea that if the lands could be brought under European-style agricultural production, with a complement of plants and animals from the Old World, then the New World could be transformed into a more productive version of the old.

This belief persisted for centuries, evidenced by the colonial practices of land expropriation for the purposes of settlement, farming, and resource extraction. These strategies are still being employed in many parts of the Americas today. A case in point comes from the colonially inspired society where I live—British Columbia, Canada. Beginning in the mid-1820s and accelerating in the 1850s, Europeans claimed, in the name of the British Crown, virtually all of the lands and resources of the aboriginal peoples in the region. They justified this takeover by noting that the original inhabitants (people they freely called savages) were not engaged in what they considered to be useful and productive agricultural activity, as defined by nineteenth-century concepts of what constituted agriculture. In some fundamental sense it was believed that this justified allowing immigrants, mostly recently arrived from Europe and the United States, to take over indigenous peoples' lands and resources. Europeans considered agriculture to be the *condicio sine qua non* of civilization. Its absence disqualified peoples from land use and territorial claims—even those that had existed for thousands of years.

The colonial governments and administrators who held this view of agriculture's power to discriminate between those who were productive (and therefore could be owners of land) and those who were not (and therefore could be dispossessed of their lands and resources) were understandably interested in almost all aspects of agriculture—although, unlike present-day archaeologists and botanists, seldom were they interested in questions of agriculture's origins. Often it was enough to explain Western agriculture as a divinely inspired gift: God caused "the grass to grow for the livestock and plants for man to cultivate, that he may bring forth food from the earth."[17] Not only did God provide domesticated animals and plants, but, when people needed a refuge, the Lord gave that as well: "For the LORD thy God bringeth thee into a good land, a land of brooks of water, of fountains and depths that spring out of valleys and hills; A land of wheat, and barley, and vines, and fig trees, and pomegranates; a land of oil olive, and honey; A land wherein thou shalt eat bread without scarceness, thou shalt not lack any thing in it" (Deuteronomy 8:7–9).

For untold generations of people who were educated in the Judeo-Christian tradition, passages such as this one from Deuteronomy connected people to God through fertile land and its agricultural bounty—with wheat and barley, the two main grain crops mentioned, and the bread that they provided inevitably listed first among the earth's riches. This is so important in Western thought that it permeates our view of what type of land is best (that which yieldeth grain), what foods are best (that which cometh from grain), and what labors are best (those which produceth grain and bread).

However, the Age of Enlightenment, beginning in seventeenth-century Europe, sparked a shift toward scientific explanations for the origins of agriculture. Still, the primacy of Judeo-Christian conceptions of agriculture is evident in the way Western philosophers and early anthropologists and archaeologists framed their questions about how agriculture and food production came about. To answer them, scholars turned to historical disciplines in search of ways to gather information of an empirical nature. The discipline of archaeology became one of the few avenues for exploring agriculture's origins because, in the absence of historical documents describing this ancient process, archaeologists were able to develop methods for peering into the past by way of the remnants and residues left behind. It must be recognized, though, that the initial foundations of this interest were predicated on the notion that agriculture holds a privileged place in Western origin stories.

In the past fifty to sixty years, there have been many attempts to summarize the history of theorizing about agricultural origins. However, with a few notable exceptions, most explanations of agriculture's origins either focused on Old World agriculture or used it as a baseline for their arguments. Yet not all of Old World agriculture tends to be of interest to archaeologists—just the complex of very early plant and animal domestication in the Near East.[18] The evidence for plant and animal domestication in that region provides a formidable connection between the rise of some of the world's earliest civilizations and the origins of the crops and animals that dominate today's agro-industrial economies: wheat, barley, oats, cattle, sheep, and goats.

The focus on this region and these domesticates provides a compelling story—particularly to peoples of European ancestry, who trace their lineages to the Indo-European-speaking forebears who are hypothesized to have done the following. Around 7,000 years ago, farming people had domesticated the above-mentioned suite of plants and animals in the "cradle of civilization"—that is, the Fertile Crescent. Flushed with the successes of agricultural plenty, they outproduced (and outreproduced) their neighbors and began to stretch their legs. They expanded in all directions over the next 4,000 years, more like a flood than a glacial advance, eventually lapping up against nonagricultural peoples and natural environments that were resistant to their advances (for the time being). The evidence for this basic story consists of many independent lines of research and observation and has been most clearly articulated by renowned archaeologists Peter Bellwood and Colin Renfrew.[19] In addition to archaeological evidence (such as house remains, pottery, tools, and plant and animal remains), considerable linguistic and genetic evidence has accrued, much of it supporting the Neolithic period expansion hypothesis of Indo-European-speaking farmers moving "out of the Fertile Crescent."[20]

This story, and the explosion of research that has generated it (and been generated by it), illustrates the theoretical importance and role of agriculture in defining the essence of "civilization." It shows how early

Near Eastern, South and Southwest Asian, and, eventually, European civilizations were essentially farming societies. It also shows that these were a particular sort of farmers who relied on a combination of small-grained cereal crops that could provide food and drink and animals that could provide meat, milk, fibers, and leather. But it has long been known that there were other sorts of farmers as well: the "barbarous" agriculturalist, horticulturalist, and pastoralist peoples—as early anthropologists would have labeled them—who were without civilization (or , at most, possessed what were once considered "archaic civilizations"). These peoples inhabited the New World.

MAIZE AND THE EVOLUTION OF CIVILIZATION IN THE NEW WORLD

When European explorers, conquistadors, and colonists first encountered the peoples of the Caribbean and the eastern coasts of the New World, from North to South America, they invariably remarked on the wonderful new foods they were offered. In the earliest such encounter, in 1492, Christopher Columbus, a magnificent navigator but a rather inept botanist, repeatedly wrote in his log about the wondrous plants of all types that he and his crew saw. He invariably compared them with the Old World plants that they knew, and so, for example, agave (*Agave* sp.) became aloe (*Aloe vera*), sweet potatoes (*Ipomoea batatas*) became yams (*Dioscorea* sp.), and maize (*Zea mays*) became millet (various genera of the Paniceae tribe). During his first voyage he discovered the islands of the Bahamas, Cuba, Hispaniola, and a few others before returning to Spain. In the daily journal he kept during this expedition, which was the first of four trips, one is struck by the almost complete absence of reference to maize. The people he met brought him manioc (*Manihot esculenta)* and sweet potatoes (both of which he referred to as yams—a plant familiar from Africa), and he noted many times that these were the foods that formed the mainstay of the native peoples' diets. Yet, with an eye to future funding for his exploration project, and knowing that he would send his diary to the Spanish royal court of Isabella and Ferdinand, Columbus wrote of the fertility of the land and described how large expanses of it could be planted with grain. But his interest was clearly not food. Columbus was looking for ready sources of wealth—commercially important goods that could be easily transported: precious metals, pearls, spices, and industrial plants such as cotton, pinewood, and mastic (a resin which was used for shipbuilding).

Within eight years of Columbus's first landfall on what are now the islands of the Bahamas, other explorers had discovered maize. We don't know exactly how maize first made its way back to the Old World, but within perhaps ten years of the Columbian Encounter it had been transported across the Atlantic and was introduced to Europe, Asia, and Africa. Portuguese explorers may have been the first to bring maize to Africa as they expanded their trade between Europe and New World. In his examination of the impact of maize on Africa over the past five centuries, historian James McCann argues that there is no clear record of maize's earliest dispersal to Africa and that the first historical document to describe it comes from a Portuguese pilot who, in 1540, said that maize was planted on the Cape Verde islands and all along the coast of West Africa.[21]

But long before maize appeared in the land of the Europeans, Europeans appeared in the land of maize: Mesoamerica. Intermittent contact between Spanish explorers and the peoples of the Caribbean coast of Veracruz and Yucatán between 1492 and 1518 alerted the Spaniards to the treasures of Mexico. A quarter century after Columbus's first journey to the Americas the indomitable conqueror Hernán Cortés began to make his way into the interior highlands of Mexico and the heart of the Aztec empire, and he was no more interested in maize than his predecessor had been. Cortés and his conquistadors were keen to find gold and other riches, but they still had to eat. Once they left Cuba and the other Caribbean islands, where manioc (also called cassava or yuca) was the staple food crop, they discovered that the peoples of the mainland relied much more heavily on maize than on any other food. All along the coast the travelers noted maize fields and plantations, some near houses and others far from towns.[22] To European eyes maize appeared to be the New World counterpart of Old World grains, such as wheat and barley (figure 2.3).

By 1519, when Hernán Cortés began his march inland from Veracruz toward the Aztec capital of Tenochtitlán, it became clear not only that maize was a crucial staple for the indigenous people but that it would become the conquerors' main food as well. Everywhere the conquistadors traveled in Mexico and Central America they came across fields sown with maize and other crops. This was of vital interest to the Spaniards because, unlike modern armies with their MREs, the conquistadors and their native allies lived off the land.

Bernal Díaz, one of the conquistadors who was with Cortés, wrote a firsthand account of these travels in stirring prose that, although he was

FIGURE 2.3. View toward the location of Villa Rica on the Caribbean coast, taken from the archaeological site of Quiahuiztlán in Veracruz, Mexico. Villa Rica is where Hernán Cortés and his conquistadors, who had allied with the local Totonac people, built their first settlement prior to their march inland. (Photograph by Gengiskanhg, Wikimedia Commons)

describing events in which he had participated sixty years earlier, as a young man in his early twenties, still has the power to captivate our imaginations today. His vivid history of the conquest is indeed "a most remarkable one."[23] He often writes of Cortés's strategy to supply his conquistadors with food procured along the way. For example, when the expedition landed at San Juan de Ulúa (then a small island forming the port of Veracruz, Mexico) during Easter 1519, they soon began to run out of the food and supplies they had brought with them from Cuba. Díaz writes: "We had no provisions, our cassava bread was getting short and mouldy, and some of the soldiers . . . were sighing to go home."[24] Cortés, of course, had no intention of returning to Cuba. According to Díaz, the commander said, "It would be as well to see what this land contained, and in the meantime to live on maize and such provisions as the Indians had in the neighbouring towns, which would give us enough to eat, if our hands had not lost their cunning."[25]

During the next many months and years the local supplies, produced and prepared by native farmers, sustained the Conquest (figure 2.4). Indians brought the conquerors maize-cakes (*tortillas*) along with other

FIGURE 2.4. A fragment of Diego Rivera's famous Huaxtec Civilization or "Maize" mural in the National Palace in Mexico City (1950). Men tend the fields in which maize plants are growing while women are shown in the foreground preparing maize dishes: grinding maize with a mano and metate, boiling maize, making tortillas and cooking them on a griddle, and making *atole*—a maize gruel. (Photograph by O. Mustafin, Wikimedia Commons)

foods and drinks. For example, Díaz writes that at Tlaxcala, "chiefs arrived with a great supply of poultry and maize-cakes and prickly pears and the vegetables that grew in their country. The camp was now very liberally supplied, and in the twenty days that we stayed there we always had more than enough to eat."[26] Some four years later, describing the march into Chiapas with Captain Luis Marin, Díaz recounted that the Chiapanecs living near Ixtapa had "large plantations of maize and other vegetables," and that when the Spaniards entered the deserted town they found that "there was much maize and other supplies there and we had plenty to eat for our supper."[27]

Maize was essential; without it people would have starved. The link between the large populations of the kingdoms of Mesoamerica and the foods that sustained them was clear. Without the primary crops, chief among them maize, the populous societies in many parts of the New World would not have existed for long. This connection makes maize a prime variable in explanations for how Mesoamerican and many other of the New World's complex societies with large populations came about. Many foods could probably have been taken out of the dietary equation— but maize could only be removed at the risk of population collapse. The overriding significance of maize as a staple food crop in the largest, most complex polities in the New World at the time of the arrival of Europeans dovetails nicely with explanations that highlight agriculture's role—particularly grain agriculture—in the rise of Old World civilizations.

However, the foregoing historical sketch of the initial European encounter with maize gives rise to a fascinating paradox. On the one hand, Europeans immediately noticed that the peoples of the New World generally lacked agricultural practices that they considered to be hallmarks of Western civilization—especially wheat cultivation and animal husbandry—a point clearly captured by the previously cited quotation from Adam Smith's *Wealth of Nations*. On the other hand, some New World crops—maize in particular—were remarkably similar to Old World grains in their productivity and storability. Therefore, if grain (i.e., wheat) agriculture nurtured Old World civilizations, then grain (i.e., maize) agriculture must also have given rise to New World civilizations. Beginning in the late nineteenth and early twentieth centuries, archaeologists who were developing global theories about the origins of agriculture and civilization found in maize enough of a parallel to Old World grain crops to single it out as the New World's equivalent to wheat. It was therefore generally assumed that maize must have had a biological and social history that was similar to that of wheat. This was a reasonable assumption at the time, but it may have led us to misunderstand the early history of maize and conflate it with the histories of Old World grain crops.

It turns out that this is a more widespread problem. We have generally neglected to theorize about the origins of non-grain crop agriculture. Although we are now accumulating information about the domestication histories of non-grain crops, we still do not have many models or explanations about how root crops, tree fruits, and other non-grain foods and spices came to be domesticated—and this goes for both Old World and New World species. There are even fewer general models about the domestication of crops that were used for non-food purposes, such as medicinal plants, or plants used for industrial purposes, including building materials, clothing, nets, cordage, and so forth.

Part of the anthropological and archaeological impetus for theorizing about the origins of food agriculture came from work aimed at understanding the growth of human populations and the rise in social and political complexity. With the emergence of ecological models of cultural evolution in the 1950s and 1960s, researchers began to look at food agriculture as part of a system-level adaptation—inextricably linking agricultural origins to broader shifts in technological, social, political, economic, and ideological organization. Agricultural production in various parts of the world was shown to be closely linked to, and in some cases even to have a causal relationship with, the processes of emerging civilization (and all that the term "civilization" entails). Since maize was, and still is, unquestionably the most widespread and productive grain crop in the New World, both in pre- and post-Conquest societies, its initial development, adoption, and spread was long seen to be important in understanding the rise of the most complex societies in the Americas. Generations of archaeologists have been raised on the proposition that maize was the most significant staple crop in the ancient New World—from the Mississippi to the Andes—even when other foods (some agriculturally produced and some not) were also known to have been of great importance.[28]

The significant contribution of maize to the pre-contact economies of many New World cultures and its overriding importance in present-day global food systems have colored our view of how maize agriculture may have originated in more ancient times. We are dazzled by maize's success, just as the early explorers were dazzled by the life-sustaining grain. Textbooks published in the late 1960s and early 1970s, such as Gordon Willey's influential *Introduction to American Archaeology,* referred frequently to the farming societies in North, Middle, and South America whose economies were characterized by the planting of reliable crops, with maize as the cornerstone.[29] The types of economic and agricultural practices embodied by maize farming were often seen as the primary means of enabling people to settle down, expand their populations, and develop arts and, eventually, political and economic complexity.

I think these deep-rooted assumptions are difficult to shake, partly because there is some truth to them and partly because our global view of the importance of grain staples meshes so well with the ideological and symbolic importance of maize in later pre-contact cultures, ranging from southern Canada to central Argentina and from the Pacific to the Atlantic coasts. But to understand the changing importance of maize, from its earliest origins to the arrival of Europeans in the New World, we need to examine those assumptions and separate final outcomes from initial motivations.

THE IMPORTANCE OF INDIGENOUS HISTORIES

Present-day indigenous peoples of the Americas tell their own histories of their origins, and these seldom mirror the nonindigenous histories produced by anthropologists, archaeologists, botanists, and geographers. In many parts of Mesoamerica, maize figures prominently in the founding histories of peoples and connects them to their land, their gods, and their ancestors. Their languages reflect these connections, containing words and stories that tell of the role of maize in the creation of the first peoples. Origin stories recount that people were oftentimes made, quite literally, of maize. Maize gave rise to ancestral beings who, with wit, intelligence, and cunning, and in concert with powerful, deified forces of nature, overcame obstacles and challenges to survive and flourish.

The Popol Vuh—the book of the ancestors of the Quiché Maya of highland Guatemala—is one of the great works of Native American literature.[30] In it, the origin of maize is laid out clearly, but metaphorically, and we see in the work a developmental sequence preserving the history of both maize's and people's origins. In the Popol Vuh, maize was first consumed as a beverage—called *boronté* by the Chortí Maya—and was one of the nine sacred drinks of Ixmucané, the god who discovered maize.[31] An animal called Hawk was said to have kneaded the blood of a serpent and a tapir into maize dough to make the flesh of the first humans.

Although the ancient roots of these histories are often shrouded and fragmentary, there is a great deal of material evidence recorded in ancient art that demonstrates the important relationship between people and maize. Various art forms show that maize was exceedingly important and that it was linked to the sacred realm—in a way that is much more profound than would be expected of a mundane food crop. This spiritual dimension of maize agriculture suggests that the grain was considered to be a food for the gods and of the gods. The reciprocal relations between people and maize mirror those between people and the world's supernatural forces.

Since maize first appeared in western Mexico, we might expect it to first show up in the art and iconography of that region. But it doesn't. So where does it initially appear? Some of the clearest iconography depicting the supernatural associations between maize and the gods was produced by the Olmec peoples of south-central Mexico, more than 3,000 years ago. We will explore this in more detail in chapter 9, which discusses how many indigenous peoples in the Americas linked maize imagery to their deities. Meanwhile, the next chapter traces the key theories of maize's origins that have been developed by the botanists and archaeologists who laid the foundations for our current understanding of this remarkable plant and the first peoples who grew it.

CHAPTER 3

Old Puzzles and New Questions about Maize's Origins and Spread

Where, when, and how was a species, so hardy that it could survive in the wild, converted to a cultivated plant so specialized and so dependent upon man's ministrations that it would soon disappear from the face of the earth if deprived of man's protection? These questions are all parts of one of the most intriguing mysteries of modern times.

—Paul C. Mangelsdorf

Before biting into your next piece of corn on the cob, take a closer look at it. You'll see an 20 centimeter-long cylinder glistening with juicy yellow kernels all attached to a stem that, if the cob is still steaming hot, provides an ideal handle. Bite into the kernels and experience an explosion of natural sweetness (but don't burn the skin off the top of your palate like I always do). This vegetable is almost too sweet to be thought of as a basic food staple. In fact, most new varieties of sweet corn seem closer to a dessert than a main vegetable course.[1] If you were to look for a wild, undomesticated version of this fruit growing on tall, thick-stemmed stalks of grass you would look in vain, for no such wild plant exists. From a distance, teosinte, the closest wild relative of maize, looks so much like maize that in fields where they grow together "even the keen eyes of the Mexican Indians cannot distinguish one from the other" (figure 3.1).[2] But after the plants flower and the inflorescences mature, it is an entirely different story. Unlike maize, teosinte has no solid cob with hundreds of grains attached to it, all tightly enclosed in a protective sheath of leaves we call the husk. Instead, an "ear" of teosinte looks like the inflorescence of many other large tropical grasses: it has from one to fifteen neighboring **spikes,**

FIGURE 3.1. A stand of maturing teosinte (*Zea mays* ssp. *mexicana*) growing in a garden in Indiana, September 2013. From a distance these plants look very much like maize. (Photograph courtesy of Fred McColly)

each about 6 centimeters long and holding from six to twelve small triangular- or trapezoidal-shaped seeds alternating along the length of the spike (figure 3.2). The whole cluster of spikes—which is the female flowering part of the plant (or more technically, the **inflorescence**)—is tended by a larger leaf, while each individual spike is enclosed by a set of smaller leaves—teosinte's equivalent of maize's husk—which dry and loosen up to let the seeds emerge upon ripening.[3]

If you were to pry the husks off the inflorescence of the two plants, the most important difference between them—apart from their size—would become immediately visible. In teosinte, the several spikes in an inflorescence are all separate, as are the individual seeds that detach from the spike when ripe, leaving them free to fall to the ground (figure 3.3). In maize, there are no individual spikes—instead, they appear to be merged into one solid unit, forming the cob, which does not break apart when ripe.[4] The maize ear is surrounded by a thick blanket of all the individual leaves that, on an ear of teosinte, would have surrounded each spike.

This is one of the staggering differences between teosinte and maize, as anyone who has ever been assigned the task of shucking a pile of corn for a late-summer corn on the cob feast can attest. But another, equally important, difference between the two is that, while maize kernels are soft and unprotected, teosinte kernels are each surrounded by a hard

FIGURE 3.2. An ear of *mexicana* teosinte with five visible spikes. Some of the individual kernels on one spike peek through the paper-thin husk leaf that partly covers it. October 2013. (Photograph courtesy of Fred McColly)

outer covering. This covering is part **rachis** and part **glume.** The rachis is the segment of the flower that, in most grasses, fastens each seed together with its neighbor and holds the whole spike together until it is ripe and ready to disperse. The glume is a hard, protective **bract** (a specialized type of leaf) that surrounds the inner part of the seed until it has fallen to the ground and can begin to germinate. In maize, the glumes are soft and, except in some rare varieties, do not cover the individual kernels, and the rachis segments are retracted from the kernel to form a major component of the cob. In teosinte, as in all grasses, when the seeds are ripe, a layer of cells between the adjoining rachis segments, called the **abscission layer,** becomes brittle, allowing the kernels to detach from one another and fall to the ground. In maize and other domesticated grasses, such as wheat, oats, and rice, the "disarticulating" mechanism found in their wild ancestors is disabled, and the adjoining rachis segments remain attached. This is what, in maize, forms the durable cob that makes it

FIGURE 3.3. Teosinte ears, bent over with kernels fully exposed, November 2014. The ear on the right side shows how the seeds, when matured and hardened, could naturally separate from one another and fall to the ground, where they could sprout during the next growing season. (Photograph courtesy of Fred McColly)

possible for farmers to gather the entire group of seeds intact, allowing them to avoid having to get down on their hands and knees to pick up thousands of disarticulated maize kernels at every harvest.[5]

The puzzle of how maize came to be so dramatically different from teosinte, or any other wild grass for that matter, has spurred the curiosity of botanists, farmers, and even archaeologists for generations. The differences between maize and the members of the group of wild teosintes that could have been its ancestors have been especially perplexing because none of the other major cereal crops that have been domesticated have undergone such massive changes from their ancestral wild forms. Wheat, oats, barley, rice, millet, and other major cereal crops all have recognizable wild cousins that, to a certain extent, can be harvested in the wild today. From a farmer's point of view, the greatest improvements to these grasses during early domestication were the selection of "non-disarticulating" rachises and increased grain size. None of these other grasses developed monstrous ears with enormous cobs that formed a solid backbone on which up to twenty rows of kernels could be lined up side by side and packaged together under one all-encompassing protective husk. Furthermore, none of these other important cereals lost an outer layer that was as hard and tough as teosinte's. In teosinte, this protective layer is what makes it possible for the ripe seeds to be consumed by grazers, such as cattle, and pass through their digestive tracts relatively unscathed.[6] In fact, this is one of the other key puzzles about teosinte, and it led eminent botanists such as Paul Mangelsdorf and Robert Reeves to doubt that teosinte alone could have been the ancestor of maize. Even when fully ripe, teosinte is all but inedible to humans, and it is difficult to extract the nutritious grain from its protective layer of tough rachis and indurated glume. The striking dissimilarity between maize and teosinte made some botanists look elsewhere for the ancestor of maize, while others have tried to explain the kinds of genetic changes necessary to produce maize from teosinte.

We will look at one of the key debates that shaped the research on maize's origins during the last two-thirds of the twentieth century and still reverberates in recent publications and studies about maize today. Of course, this interest in maize and the debates that the plant has generated are not just esoteric musings. Maize has been the source of such controversy and debate because of its pivotal role in today's global food supply. Understanding its origins and the history of its domestication has implications for our understanding of how it can be modified and manipulated to be more productive, more drought and pest resistant, more variable in its desired qualities and products, and so on. Literally billions of dollars, millions of jobs, and ultimately tens of millions of lives are at stake. Knowing as much as possible about maize, one of our primary food sources, may be key to the survival of our children, our grandchildren, and many generations to come—just as it was vital to the survival of the Mesoamerican peoples who, we now know, first domesticated teosinte to create maize.

SEARCHING FOR THE ANCESTOR OF MAIZE

Recently I managed to buy a secondhand copy of the late Paul C. Mangelsdorf's *Corn: Its Origin, Evolution and Improvement*.[7] Published in 1974, this book summarizes a lifetime's study of maize by the famed

Harvard botanist, who virtually defined the field of archaeobotanical maize studies. Inside the cover of my prized volume, Paul Mangelsdorf inscribed a dedication: "To Jim Cameron: My first graduate student at Harvard." Cameron (not to be confused with James Cameron, the director of the blockbuster movies *Titanic* and *Avatar!*) and Mangelsdorf published several important papers about maize together in the 1940s.[8]

In *Corn,* Paul Mangelsdorf describes how, as a young botanist in the 1920s, he came to be interested in the origins of maize and was inspired to join in the search for its ancestor. Through planting experiments and detailed botanical analyses he and his colleagues and students hypothesized that modern maize descended from a now-extinct wild maize ancestor—pod corn. Although he didn't explain where wild maize originated, he proposed that it hybridized (crossed) with a sister grass, a genus called *Tripsacum (Tripsacum dactyloides),* to produce a subspecies of teosinte (*Zea mays* ssp. *mexicana*). Modern corn, according to his hypothesis, was the result of further crossing of pod corn with teosinte or *Tripsacum* or both.

Mangelsdorf's book summarized most of the archaeological and botanical information that was known about maize in the early 1970s, and, in many ways, it was a staunch defense of his hypothesis, called the "tripartite" hypothesis for its three distinct propositions (although we could equally call it that after the three amigos of his model: wild maize, teosinte, and *Tripsacum*). The three parts of the tripartite hypothesis were:

(a) cultivated maize originated from pod corn [of South American origin],
(b) teosinte is a derivative of a hybrid of maize and *Tripsacum,*
(c) the majority of modern corn varieties are the product of admixture with teosinte or *Tripsacum* or both.

This hypothesis, first published in the late 1930s with R. G. Reeves, was elaborated in many subsequent publications and rose to become the orthodox view of maize domestication for several decades.[9]

Right from the beginning, though, it met with resistance. In 1939, another botanist and geneticist, George Beadle, building on work published in 1875 by the German botanist Paul Ascherson, along with other earlier research, elaborated a different hypothesis, which suggested that all modern maize was descended from one of the teosintes.[10] His reaction to Mangelsdorf and Reeves's proposal is simply put: "This supposition has no obvious advantages and appears to be unnecessary."[11] Beadle's model was elegant in its brevity and depended on no mysterious extinct wild ancestors. He had been studying *Zea-Euchlaena* hybrids since the early 1930s (teosinte was classified as *Euchlaena mexicana* Schräder before it was moved over to the genus *Zea* in the 1970s) and was convinced that teosinte must have been the wild ancestor of maize. Furthermore, in his 1939 rejoinder to Mangelsdorf and Reeves he spelled out the five main genetic mutations that must have been selected for by ancient farmers.[12] Beadle also suggested that the first ancient Mesoamerican peoples to use teosinte may have been interested in the stony grains as a type of popcorn or, perhaps more appropriately, "popteosinte." By popping kernels of teosinte during many of his public lectures on the topic, Beadle demonstrated that the plant could have been of great interest to early hunter-gatherers since it was easily consumable with the simplest of technologies: fire. We will return to this idea later and examine the evidence for it (or lack thereof) and its implications for our understanding of the initial domestication and spread of *Zea*.

After lobbing this volley in a debate that was to percolate for the next thirty years, George Beadle abandoned maize for research on fungus. He went on to study the genetics of *Neurospora crassa*—a variety of red-colored bread mold—eventually winning a Nobel Prize for his discoveries in 1958. His research and administrative duties (including a seven-year stint as president and chancellor of the University of Chicago that lasted until his retirement in 1968) prevented him from returning to his teosinte interest until the early 1970s.[13] During the interim, from the 1940s to 1960s, Beadle's view of maize's origins had seemingly been discredited by Mangelsdorf's analyses of archaeological collections of maize from Bat Cave in New Mexico, several caves in the state of Tamaulipas, Mexico, and the vast assemblage of maize discovered by archaeologist Scotty MacNeish and his colleagues in the cave sites of the Tehuacán Valley in Puebla, Mexico.

MacNeish, along with Paul Mangelsdorf and Walton Galinat, collected and studied 23,607 pieces of maize—kernels, cobs, stalks, leaves, roots, and quids (chewed pieces of maize stalk)—from the Tehuacán Valley cave excavations. Their exciting discoveries showed that the earliest maize from these deposits was estimated to be about 7,200 years old, and it was strikingly similar to the hypothesized extinct wild maize.[14] Over the next decade a series of publications, including the five-volume report on the Tehuacán Valley excavations and Paul Mangelsdorf's previously mentioned book, *Corn*, seemed to buttress the tripartite hypothesis. But then George Beadle came back.

Between 1971 and 1973 Beadle mounted a search for maize-like "mutations" in populations of teosinte growing in the Balsas River watershed—an expedition that has been dubbed the "teosinte mutation hunt." He and his team of botanists and archaeologists were looking for any of the four or five main phenotypical traits that, while very rarely expressed in teosinte, were common in maize. Persistent selection of these traits by ancient teosinte harvesters could have led to the transformation of teosinte into maize. These main traits, governed by individual alleles, or variants, of specific genes included the shift from hard to soft glumes, from brittle to solid rachises, from single to paired **spikelets,** and from **distichous** to **polystichous** spikes. Although the group searched the fields and hill slopes of the Balsas River valley in southwestern Mexico, examining what amounted to millions of samples, they never found the holy grail of teosinte research: a wild teosinte plant with naturally occurring soft glumes (the fruitcase) and exposed kernels.[15] This mutation was a key element in the domestication process since it is one of the most obvious characteristics of maize on the cob—how would you like to bite into a cob of corn consisting of nothing but pebble-like kernels? Since teosinte seeds are individually sheathed in a durable case, it is almost impossible to extract the nutritious grain from the fully ripened ear without a great deal of labor (or a small amount of heat, as Beadle showed with his "popteosinte"). Furthermore, the individual seeds are arranged along a spike with a brittle **rachis** that shatters when they are ripe, dispersing them over the ground where, with luck, they can survive the dry season and germinate when the next rainy season begins. According to the teosinte-to-maize hypothesis, early teosinte harvesters—and eventually growers—encountered the occasional "mutant" form of maize-like teosinte that had at least some exposed—or naked—kernels, which they then selected and replanted. These plants were the progenitors of maize.[16]

Even though Beadle and his team did not find the mutation they were looking for in the wild, research proceeded, and within a few years of the teosinte mutation hunt, botanists such as Garrison Wilkes, Hugh Iltis, and John Doebley began to narrow the search for maize's teosinte ancestor even further. A few years earlier Wilkes, who had been a student of Mangelsdorf's at Harvard, published his doctoral dissertation. At that time, it was the most comprehensive study of all the teosintes. He described them in detail, with photos and drawings, providing information on their geographic distributions, habitats, and possible relationships with each other.[17] Wilkes's survey included six main groups, one of which, Balsas teosinte, grows in dense stands along the slopes of the Balsas River valley and its tributaries.[18] This was the variety that had drawn the attention of Beadle and his mutation hunters, and it was starting to look like it, along with Chalco teosinte (*Z. mays* ssp. *mexicana*), were the prime candidates for being maize's wild ancestor.

Based on their field and lab research in the 1970s, the botanists Hugh Iltis and John Doebley reclassified the genus *Zea* into four species: *Z. mays* (with three subspecies: *mays, mexicana,* and *parviglumis*—all annual teosintes), *Z. luxurians* (also an annual), and *Z. perennis and Z. diploperennis* (both perennials). Balsas teosinte (*Z. mays* ssp. *parviglumis*) and Chalco teosinte (*Z. mays* ssp. *mexicana*) are both morphologically and genetically closest to maize (*Z. mays* ssp. *mays*), and therefore all three were classified as the same species, while other varieties, especially the perennials, were labeled as different species.[19]

Around this time, in 1980, Beadle published a paper in *Scientific American* arguing that teosinte was likely the ancestor of maize. He noted that, with a handful of genetic changes of large effect, this annual teosinte could have transformed into an early version of modern maize.[20] However, new techniques and methods in the study of plant genetics were starting to have an accelerating impact on maize research, especially in providing new perspectives on maize's origins. A few years after George Beadle's paper appeared, John Doebley and his graduate and postdoctoral students and colleagues began to publish an impressive body of genetic research that helped weed out some of the alternative hypotheses that had dominated the maize debate during the previous forty years. In one key study, Doebley and his colleagues showed that the chloroplast DNA of domesticated maize was most similar to that of *Z. mays* ssp. *parviglumis*.[21] This finding was based on analysis of **isozymes** (enzymes produced at the molecular level) and was much more definitive than previous types of genetic analyses that had allowed only an assessment of the similarities and differences between the numbers and shapes of knobs on the stained chromosomes. The isozyme analysis allowed Doebley and his team to conclude that *parviglumis* was likely the progenitor of maize, because its isozymes matched maize samples more closely than any of the other teosintes. As we will see in chapter 8, ongoing refinements in genetic analyses have permitted even more robust comparisons among the varieties of *Zea* and its closest cousins, and they have confirmed the hypothesis that modern maize is primarily descended from Balsas teosinte (*Z. mays* ssp. *parviglumis*).

Although Mangelsdorf somewhat altered his tripartite hypothesis in the face of mounting evidence in support of the teosinte-to-maize hypothesis, the new genetic research has, for most botanists, laid his hypothesis to rest.[22] Even so, there have been some more recent attempts to revive, or at least modify, the tripartite hypothesis, but these have been met with stiff opposition.[23] Most of the geneticists studying maize and its close family members now see no possibility that any plant other than an annual teosinte (*parviglumis* especially and, to a somewhat lesser extent, *mexicana*) could have given rise to maize.

At the same time the new genetic studies have opened the door to a series of reinterpretations and new discoveries about maize's dispersal. For example, one theory that kept popping up until the establishment of *Z. mays* ssp. *parviglumis* as maize's primary ancestor was the notion that maize may have had two or more original centers of domestication—at least one in Mesoamerica and one in South America. Paul Mangelsdorf's view on the matter was that certain varieties of maize, particularly the popcorns, were so characteristic of South American varieties that the wild ancestor of some maize lineages could well have been a now-extinct South American popcorn. In fact, in his 1974 book, Mangelsdorf postulated that the more than one hundred races of maize could be divided into six lineages, four of which were of either Colombian or Peruvian origin, and only two of which were Mexican.[24] This idea was certainly reasonable if one thought that maize descended from an extinct wild maize instead of from teosinte (no species of which are native to South America). Of course, for geneticists such as George Beadle, who had, since the 1930s, concluded that maize was actually a domesticated form of teosinte, there was little likelihood that maize could have emerged anywhere but Mexico.

The idea that maize could have been independently domesticated outside of Mesoamerica was plausible only as long as its origin was not linked to teosinte. In a fascinating chapter in his 1974 *Corn* book, Mangelsdorf systematically debunked hypotheses that maize had been domesticated in Asia, India, Africa, the Philippines, and even Europe. Many of those who believed there were Old World, pre-Columbian varieties of maize did not think that it was domesticated in the Old World, but rather that it diffused there by way of trans-Pacific or trans-Atlantic contacts, long before Columbus first visited the New World. Mangelsdorf showed that, based on data available at the time, both scenarios were highly unlikely—primarily because there was no convincing evidence of pre-Columbian maize in any region outside of the Americas.

Mangelsdorf's idea that maize diffused within the Americas, but not outside of the Americas until European traders exported it, had some interesting implications for his theories about maize's domestication. He speculated that maize did move around. For example, he believed that a variety called Peruvian Chullpi was the progenitor of all sweet corn and that it diffused northward, eventually carrying desirable traits to the US Southwest and even farther to the northern and eastern parts of the continent. Likewise, he speculated that one lineage of maize, Pira Naranja, may have originated in Colombia and then moved southward into Chile and Argentina. So in his theorizing about maize's origins and spread, he argued that maize could move among peoples who were geographically connected to one another but could not have moved across oceans before the advent of European sailing technology.

One legacy of this theory is that maize was thought to have evolved very early in South America, certainly as early as in Mexico, and therefore that it was reasonable to think that maize could be as old there as it was in the Tehuacán Valley. We will look at the implications of this idea in chapters 4 and 5, but keep in mind that most botanists and geneticists have abandoned the idea that there were independent domestications of maize outside of Mexico. It has become clear that maize entered South America as a long-domesticated teosinte and that, while we do not know yet exactly when or how many times this took place, the molecular genetics show that it did take place. But this doesn't mean that Mangelsdorf and others were wrong about the subsequent movements of maize. Over several thousand years it is quite likely that new varieties of maize, developed under the evolutionary process of agricultural selection, moved back and forth between North and South America, as well as within the continents. In later chapters we will examine the evidence that an early form of maize was transported, traded, or moved with expanding farming populations from its initial homeland, where it was domesticated, to the other regions of the Americas—both north and south. And we will see that maize moved again and again within the Americas.

HOW WAS EARLY MAIZE SPREAD?

What accounts for the spread of maize (or any other plant for that matter) by humans? What would induce people to start cultivation in the first place and, once they were farming, to adopt new plants or animals or new varieties of old ones? This is one of the key questions in the study of agricultural origins. In the case of maize, we could ask the following sorts of questions: Why would people in Andean South America, Central America, the southwestern United States, or the northeastern United States be interested in growing a plant that they had never seen before and that may, initially at least, have had few, if any, advantages over the plants or other foods that they were already using? What would induce hunter-gatherer-fishers or farmers, living in such a wide range of environments, to start growing maize when, for thousands of years, they had managed without it?

We all know that the development of an idea is different from its spread. The question of how the teosinte plant was first domesticated—how it began its transformation into maize—involves a somewhat different set of processes from the ones that led to the spread of the already-domesticated plant. But in reality, these two processes are often interlinked in complex and unexpected ways. For example, many theorists of the origins of agriculture have proposed that one of the stages of initial domestication occurs when people move a wild plant or animal out of its natural habitat to new habitats where it would not be able to thrive on its own. Therefore, in the case of some species, the spread of a wild form outside of its homeland may end up being an initial step in the domestication process.

The development of new traits and desirable characteristics may be a result of a plant's spread, by way of "adoption," into new environmental and social circumstances. As the earliest teosinte-like maize was spread beyond its native habitats in west-central Mexico, it began to undergo significant changes in morphology and productivity, many or most of which might not have taken place if it had remained in its homeland. In turn, these changes increased the likelihood that the maize would spread farther afield as neighboring peoples saw the advantages of adopting the new plant. After a time, as maize continued to adapt to an ever-increasing array of new environments (both natural and social), new varieties may have moved back into the regions where it first appeared, replacing the initial domesticated variants. Over the course of hundreds or thousands of years it is possible for waves of new varieties to sweep across regions, replacing or mixing with older varieties. It is also possible that characteristics of interest in later times were quite different from those that were of initial interest. So, for example, people may have initially adopted maize in one region because it was a reliable source of green, nutritious vegetables at a time of the year when such foods were otherwise scarce. Later on new varieties with larger ears and more grain per ear may have been adopted because they provided storable carbohydrates. Over the course of many generations types with even larger ears might have been imported or developed locally, and farmers could eventually have stopped using the immature green ears, instead shifting their attention to harvesting the mature dry ears. This sort of spread and changing use of the plant could have taken place without the movement of people; ideas about how to use the plant and the actual seeds themselves could have been passed from community to community through existing exchange networks among neighbors.

Another type of spread would have been the movement of people with farming practices into neighboring regions that were either unoccupied or occupied by non-farming peoples. This type of spread leaves a clearly visible footprint on the landscape and has been called the "budding-off" process of expansion—or, as mentioned earlier, "demic" expansion.[25] This is precisely the type of agricultural spread that some archaeologists are now arguing happened in Europe during the Neolithic period, between about 7,000 and 5,000 years ago. In a recent article, Peter Rowley-Conwy discusses four examples of agricultural spread where the primary mechanism, previously thought to have been gradual adoption of agriculture by Mesolithic hunter-gatherers, has now been shown to be the result of immigration by peoples who practiced a very specific kind of Neolithic agriculture, cultivating "four cereals—emmer, einkorn, free-threshing wheat, and barley—and five pulses—lentil, pea, broad bean, bitter vetch, and grass," along with four main domesticated animals: cattle, sheep, goats, and pigs.[26] As communities of successful farmers grew, their surplus population would expand ever outward, occupying new lands that were suitable for colonization by farmers—often at the expense of indigenous hunting-fishing-gathering peoples.

This hypothesis of agricultural spread (or spread of agriculturalists) by migration and colonization has been proposed for some regions in the Americas also. As in Europe during the Neolithic period, the Americas may have seen "'infiltration,' 'trickle,' or 'creep' migrations" rather than a single or repeated wave of colonizers that swept across the continents.[27] In the case of the earliest maize cultivators in the Americas, spread by colonizing agriculturalists was likely only possible after maize had become a major staple crop, or at least a significant contributor to the diet. This idea has been advocated by several archaeologists—most recently by some who have suggested that migrating maize farmers spread north out of Mexico and into what is now the southwestern United States. Combining archaeological evidence with linguistics and human genetics, researchers speculate that maize farmers migrated into Arizona and New Mexico bringing an ancestral, or "proto," version of their Uto-Aztecan language with them and in some regions displacing indigenous populations.[28] This proposal has led to a vigorous debate with other researchers who argue that the evidence is still too weak to make such claims.[29]

It can be very difficult to untangle the intermixture of these two types of spread: the adoption of domesticates and agricultural practices by neighboring populations and the wholesale movement of farming peoples into new regions. In some parts of the Americas, particularly during the initial stages of domestication, it is most likely that domesticated plants moved among peoples, whereas in later periods peoples with new or different agricultural practices may have more frequently colonized new regions. Either or both of these possibilities could have taken place repeatedly, independently, or simultaneously.

Let's look at a hypothetical scenario that illustrates some of this complexity. For example, say that some seven or eight thousand years ago mobile hunting and gathering peoples who lived on the eastern margins of the Balsas River came into contact with the descendants of the peoples who domesticated teosinte and who were growing and using the first primitive varieties of maize. These neighboring hunter-gatherers may have traded or accepted as a gift some of the early maize seeds and planted them along with some other species of plants they were already cultivating—gourds, for example. They and their descendants continued to plant maize for generations, and the plant continued to spread in similar fashion, outward from its original homeland to new environments and among different cultures. If, in one of these other regions, perhaps centuries later and hundreds of kilometers distant, people had developed a more productive variety of maize (or a variety that was more colorful, or more drought- or pest-resistant, or sweeter, etc.), the new maize could start finding its way, again through trade, back to the original group. One possibility is that this higher-quality variety of maize permitted the descendants of the original group to become more productive farmers than they otherwise would have without the new and improved maize. In fact, it might have permitted them to settle permanently and even expand their population. These increasingly maize-dependent farmers could then start expanding outward into new environments, possibly taking over the territories of hunting and gathering peoples who were still using ancestral, and possibly less productive, forms of maize. Eventually, after many generations of such expansion, farming peoples could begin to colonize entirely new regions that had previously been only sparsely populated or even unpopulated. While it moved across space and as time passed, maize would still be evolving as new uses were being found for it and as farmers helped it adapt to new environments. In this sense, maize would be both spreading and evolving at the same time. It would continue evolving (through human selection and intervention) both in its initial homeland and in all of the new distant regions where it had been transported. And it would continue spreading (through human transportation—never on its own), both internally, within zones where it was already present, especially as new varieties and traits were introduced, and externally, pushing the margins of its known territory.

Many scholars have proposed theories explaining the origins and spread of domesticated plants, including maize, throughout the Americas.[30] More often than not, their explanations and hypotheses are part of general theories of agricultural origins, some of which have been applied widely around the world. I won't review all them here but instead will look at a few theories that provide some useful clues to the factors that early peoples might have been concerned with that might have prompted them to adopt maize cultivation in the first place. Broadly speaking, conjectures about the origins of agriculture fall into two main groups: optimization theories and social motivation theories. Although they are often proposed as mutually exclusive explanations, I don't think there is any reason they can't be looked at together and combined in various ways.[31]

In his book *Guilá Naquitz: Archaic Foraging and Early Agriculture in Oaxaca, Mexico,* Kent Flannery summarizes the most prominent agricultural origin theories up to the mid-1980s. Having conducted archaeological fieldwork projects that investigated agricultural origins both in the Near East and Mesoamerica, Flannery was in a unique position to evaluate the practical implications of the then current theories and propose some new approaches. He lays out a perspective that, more than a quarter of a century later, is still the most useful that I've seen:

> The origins of agriculture involve both human intentionality and a set of underlying ecological and evolutionary principles. When we, like paleontologists, ask *how* a particular case of domestication happened, we are probably going to be confronted with a biological principle such as mutation, natural selection, or . . . coevolution. This is probably the level on which the most universal aspects of domestication lie. When we, like anthropologists, ask *why*, we may find ourselves confronted with reasons that are local and can only be modeled after we have done our best to reconstruct the local cultural pattern.[32]

Flannery goes on to summarize the various models and theories that had been relied on most heavily during the previous decades and are, in their various incarnations, still among the most commonly used today:

(1) climate change,

(2) population pressure,

(3) broad-spectrum adaptation, and

(4) coevolution.

The climate change hypothesis suggests that a period of prolonged warming at the end of the Pleistocene, beginning around 10,000 years ago, was the trigger that led people to begin interacting with plants and animals in new ways—ways that favored tending them more carefully and ultimately leading to the domestication of many species. This hypothesis requires that there be evidence of significant deterioration in climate, such as more frequent and prolonged droughts, corresponding to the time when agriculture first appeared. The population pressure hypothesis proposes that, as post-Pleistocene populations grew, people could no longer be supported by hunting, fishing, and gathering alone. This forced people in several regions of the world to intentionally intervene in the life cycles of some plants and animals, thereby beginning the process of domesticating some of the species that they had previously collected or hunted. The broad-spectrum hypothesis suggests that the shift to agriculture occurred as post-Pleistocene peoples began to exploit a much wider range of plants and animals than they had during the Pleistocene ice ages. In the aftermath of the last ice age, people began to harvest species of plants and animals that had hitherto been largely unavailable or ignored. They may have begun to rely on these new resources as a result of the decreasing availability of large game animals and other easy to obtain foods as human populations grew. Although many of these new or previously ignored foods, such as cereals, fruits, and root crops, came in smaller "packets," which required more energy to harvest and process, they were much more plentiful, reliable, and storable than foods derived from larger species, such as big-game animals. By broadening the spectrum of foods at their disposal, post-Pleistocene peoples put themselves in contact with, and began to more frequently rely on, several species that were amenable to domestication and would eventually form the basis of all subsequent agricultural economies. Finally, the coevolution hypothesis, proposed by David Rindos in the early 1980s, suggests that humans and particular species of plants and animals evolved together—and that over a very long time their interactions conferred evolutionary advantages on one another. As envisioned by Rindos, this process was unintentional and gradually led to domestication and then to the practice of agriculture. Unlike the previous three hypotheses, the coevolutionary hypothesis doesn't have a specific trigger—instead, it assumes that domestication followed by agriculture is the inevitable evolutionary outcome of human-plant (and some animal) interaction.

Elements of all of these hypotheses do make sense, especially if one doesn't look too closely at the details. Inevitably, though, the general applicability of these hypotheses starts to break down when we try to apply them to particular places and time periods. So, for example, while the population pressure hypothesis might explain why people began to expand the range of foods they used in some regions, it doesn't in other regions. Flannery points out that population pressure isn't a very convincing trigger for the development of agriculture in the Oaxaca Valley in Mexico, because population remained at very low levels both during and after the several-thousand-year-long stretch of time when key plants were domesticated and people adopted farming.

Flannery found that the most useful approach for understanding agricultural origins among the ancestors of Oaxaca's Zapotec peoples is one that combines both the general-universal models and the particular-local aspects of emerging agriculture. To explore the process of emerging agriculture in Oaxaca and understand the interactions among a wide range of factors that may have influenced it, Flannery and his colleague Robert Reynolds developed a computer simulation program to look at the long-term impacts of all of these factors on the decisions that the ancient Oaxaqueños may have made during the period corresponding to their occupation of Guilá Naquitz cave and the surrounding sites in the region. By modifying variables such as seasonal and long-term changes in rainfall, population reproduction rates, plant and animal resource choices, and memory of past decisions and their effectiveness, Reynolds and Flannery were able to model how a transition from hunting-and-gathering to agricultural economies would look. The strength of this approach is that it predicted similar changes to those actually observed in the archaeological record at Guilá Naquitz. This simulation model was based on particular local knowledge of resources and conditions, but it also incorporated general processes, such as reproduction rates, nutritional impacts, and so forth.

Flannery and Reynolds, along with Rindos a few years earlier, elaborated, formalized, and tested the idea that agriculture most likely developed as a response by hunting and gathering peoples to buffer against the inevitable seasonal and yearly environmental variations in resource availability. Both present-day agriculturalists and their early ancestors were interested in making sure the plants and animals they relied on were available over the long run, during both times of plenty and times of scarcity.[33] Flannery and Reynolds proposed that people living in the Oaxaca Valley could have tended plants of importance to them (not just food staples, but spices and medicines also) so that they would have them available even during drought years, when there was very little chance they would grow on their own. This strategy is, in fact, frequently used by hunting, gathering, and fishing peoples all over the world who make their living capturing or harvesting non-domesticated resources but who also, from time to time, engage in cultivation or animal husbandry. Eventually, after generations of such interventions, the plants undergo genetic modifications in response to the selective pressures imposed on them by human attention. For example, if seeds are kept for replanting, then the selective pressures that would operate in the wild are truncated and new varieties are allowed to survive under human care. This critical intervention in the life cycle of the plant would be encouraged by the increased chances of survival of the human population. This theory is particularly important for food plants, especially when those plants provide significant nutritional and caloric additions to the diet. It is not so easy to see how this process would work for species that were used as supplements rather than significant sources of food.

This brings us to the social motivation theories. Some proponents of these theories suggest that hunting, gathering, and fishing peoples began to develop social structures that favored novelty and innovation. In social environments that encouraged competition for status and prestige, it is possible to imagine a scenario where groups wanted to produce more plants and animals that could be used in feasting or social display. Brian Hayden has proposed that this mechanism was one of the ways that agriculture was initially developed, adopted, and spread.[34] The basic idea is that people's interest in plants and animals goes far beyond simply using them as food for survival. Although survival is, of course, an abiding interest, there are additional concerns generated by the social nature of human existence. As people engaged in social rituals that involved food exchanges, feasts, and displays, there would emerge demands for increasing production as well as novelty—especially for highly valued foods and beverages. This set of social factors would reinforce the desire to put more effort into the production of foods that would be useful for both display and survival. These two goals are not mutually exclusive and, in fact, could be mutually reinforcing. There is growing evidence that maize was subject to both types of selective pressures from humans at various times in its history and in various places.

In recent years archaeologists have begun to observe that there are two very distinct stages of maize adoption and spread. The first stage is the early movement of maize into a wide range of environments extending northward to northern Mexico and the southwestern United States and southward into the lowland tropics of Central and South America. In Arizona and New Mexico, for example, maize was present by at least 4,000 years ago, especially in the southern regions of the Southwest. Farther north, researchers have found that maize was also present among mobile foraging peoples in the Late Archaic period but that it was not a major dietary staple until about 1,000 years later, when the Basketmaker II people arrived.[35] Maize was therefore known and used in ways that we don't yet fully understand for perhaps 1,000 years or more before it became an essential part of the diet by about 3,300 years ago.

In some regions of lowland Mesoamerica it looks like maize was present among Archaic period peoples long before they lived in permanent, year-round villages. Along the southern Pacific coasts of Mexico and Guatemala, maize pollen and phytoliths have frequently turned up in stratified deposits of lagoon sediments. Likewise, these types of maize microremains have been reported in many contexts along the Gulf coasts of Mexico and Belize. In one case they are estimated to date to 7,300 years old. This suggests that for several thousands of years, until the advent of settled village life, mobile Archaic period foragers and fishing peoples were growing some crops, including maize. By 7,800 years ago these peoples may have moved or traded maize as far south as Ecuador, where maize pollen and phytoliths have been found in some of the earliest settled villages in the Americas.

In both cases—in the southwestern United States to the north and the Pacific and Gulf coasts of Central and South America to the south—mobile peoples, and even early sedentary peoples, may have been growing maize as a supplemental food, while their descendants may have had different uses for it. There is very little evidence that early maize was a staple food crop anywhere until 3,000 to 4,000 years ago. This doesn't mean that it wasn't a food, just that its contribution to the diet was far less than in later times (after 3,000 years ago) when maize had become much more like the large-eared plant we know today.

This period marks the beginning of the second stage of maize use—a new wave of expansion and spread that has continued up to the present day. This second stage is marked by the development of new varieties of maize that had increased productivity and were fully adapted to the new conditions of the environments where they had been transported. This idea was first developed by Ann Kirkby while she was working for the Oaxaca archaeology project in the 1960s and 1970s. She, along with Kent Flannery and the Oaxaca archaeology team, proposed that settled village life in Oaxaca became possible when people had access to new varieties of maize that were larger and more productive than previous varieties. This is an interesting case because Oaxaca is also the location of the earliest maize cobs, which date to about 6,200 years old. Maize was clearly available and being used by Archaic period foragers in Oaxaca at this time. It was also being grown and eaten by the first villagers whose permanent settlements appeared about 2,000 years later (ca. 4,000 years ago). But more intensive farming really started when the descendants of those first villagers began to rely on maize as a food staple, about 1,000 years later. This is when annual maize crops could be counted on to yield more edible grain per unit area of cultivated land—likely as a result of genetic changes to the plant brought about by the farmers' diligent selection of increasingly productive varieties, as well as the use of more intensive agricultural practices.[36]

Farther south, in Chiapas, my colleagues and I found a similar pattern. Maize was present during the Archaic period, and it was even very common in the earliest villages, which dated to the Early Formative period and were contemporary with those just mentioned in Oaxaca. But even though maize was present in early villages since their inception around 4,000 years ago, it did not register as a dietary staple until a thousand years later—around 3,000 years ago—a pattern that increasingly appears to be the case over much of the Americas, where maize was first used as a food staple. Unlike in the US Southwest, where it seems likely that Basketmaker people immigrated into the region, bringing their more intensive maize agriculture with them and replacing the Archaic foragers who only sporadically used maize, the Early Formative peoples in Chiapas and Guatemala may have developed out of the preceding Archaic cultures and, at least initially, continued using maize as a supplement rather than as a staple. New, more productive varieties of maize may have been developed or introduced, as they had been in Oaxaca, and this could have helped spur on increased reliance on maize agriculture. As we will see in chapter 7, this change is clearly marked in the chemical signatures of the bones of ancient villagers.

As with the Basketmaker example in the US Southwest, the Formative period villagers of southern Mesoamerica began to spread out, settle in new areas, and increase in population, displaying a cultural similarity over a broad region. This same pattern may have taken place even earlier with the Valdivia peoples of coastal Ecuador. Based on the ubiquitous presence of microremains of maize (phytoliths and pollen) in Valdivia and pre-Valdivia period settlements, villagers were certainly maize (and other plant) growers, but they may not have actually depended on the crop as their staple food. They do not show the characteristic chemical signature of maize consumers that we see in later time periods.

Why, then, were they growing maize if they weren't relying on it for food? What role did it have in their diet? How was maize integrated with other plant cultivation and with fishing, hunting, and the gathering of wild species of plants? These key questions arise out of the puzzle of maize's initial domestication and spread because it is clear that the obvious explanations (for example, that they grew it to supply a vital food source to support village populations) cannot have been the case, since maize was domesticated and spread during its first several thousand years of use by people who apparently did not eat much of it and who, for the most part, were not sedentary villagers. Later, during the first millennium of settled village life, people still did not initially rely heavily on maize. It was just one of many foods that people used and not necessarily the most important. In the next chapters we will take a closer look at the timing of maize domestication and spread, along with new evidence for maize's shifting roles in ancient societies.

CHAPTER 4

Timing Is Everything

Dating Maize

On July 19, 1835, the HMS *Beagle* anchored in the seaport of Callao on the coast of Peru. Several times during the ship's six-week stay in Lima, Charles Darwin disembarked from one of the *Beagle*'s small boats to clamber along the shoreline of a smallish, barren-looking island named San Lorenzo, within easy sight of the mainland. He must have been happy for the break from Callao, which he describes thus: "Callao is a filthy, ill-built, small seaport. The inhabitants, both here and at Lima, present every imaginable shade of mixture, between European, Negro, and Indian blood. They appear a depraved, drunken set of people. The atmosphere is loaded with foul smells, and that peculiar one, which may be perceived in almost every town within the tropics, was here very strong."[1]

Fortunately for us, Darwin spent some time exploring San Lorenzo, where he enjoyed relative security compared with the rough and tumble seaport, because it was here that he made one of the earliest discoveries of archaeological maize. Darwin was puzzled by the appearance of beds of seashells many feet above the present-day ocean level and wondered if the land had been uplifted as a result of some geological process in the distant past. He speculated that the time span to produce such an effect must have been great, and this was one of the countless observations he made that helped him see the link between time and biological change—observations that were to have a momentous impact on the development of modern science. One thing, however, didn't seem to fit with his geological interpretation of the shell layers: "On the Island of San Lorenzo ... I was much interested by finding on the terrace, at the height of eighty-five feet, embedded amidst the shells and much sea-drifted rubbish, some bits of cotton thread, plaited rush, and the head of a stalk of Indian corn. I compared these relics with similar ones taken out of the Huacas, or old Peruvian tombs, and found them identical in appearance."[2]

IT'S ALL RELATIVE: DATING BY ASSOCIATION

Just as Darwin observed some 180 years ago, the age of ancient objects, whether they are artifacts made by people or the remains of long-dead organisms, can be estimated by their association with other objects or features of known age in the environment. In the case of San Lorenzo Island, Darwin allowed for the possibility that the shells (which he may have assumed were natural deposits rather than the remains of ancient meals of seafood) sat on a terrace that rose, through ancient geological processes, above the sea level. He thought that the artifacts that he noted mixed in with the shells had washed in before the terrace was formed. But—and this is an important link—he noticed that these artifacts were "identical" to others that had been observed in ruins along the Peruvian coast. Darwin was, in fact, dating the shell layer by association—a practice that is essential to observing the passage of time in both archaeology and geology.

Darwin's brush with Andean archaeology is particularly fascinating because he was one of the first people to link the landscape, artifacts (such as cotton textiles, rush matting, and pottery), and subsistence remains (shellfish and maize) by stratigraphic association. Furthermore, he noted something that, while rather common along the arid coast of Peru, is extremely rare in the rest of the New World: the perseveration of plant materials in open-air sites.[3] For archaeologists, though, his observation of maize in association with artifacts similar to ones found in other ancient sites is of singular interest. Is it possible to assume that this association is the result of action by people, or is it the result of some natural process (in the case of the San Lorenzo Island materials, sea-drifted rubbish)? Darwin did not assume that the maize, other artifacts, and shells were all the same age. They were associated with one another at the time he observed them, but the artifacts could have drifted in on the tide—presumably when the shell deposits were still at sea level.

Present-day archaeologists might look at the same deposit and offer a different interpretation. Perhaps the maize, other artifacts, and shell-fish remains had been deposited as refuse in a heap spread over the terrace (such deposits are called **middens** by archaeologists). Their association would therefore be a result of cultural activity rather than natural processes. Another possibility is that ancient peoples brought the materials from a distant location to provide construction fill to build up a raised platform, such as the base of a pyramid, temple, or other monument. Called *huacas* in the Andean region, these human-made features (as well as some natural landforms) are considered to be sacred by the indigenous peoples of the area.

Darwin's interpretation in this case is illustrative of the difficulty inherent in the dating of materials by association. The shells and the maize may have been the same age, if they were deposited together as the contents of a midden. Or, they may have been of quite different ages if the maize and other artifacts washed in or were somehow deposited at a later date. As we will see, estimating the age of maize domestication and spread has been fraught with the same sort of problems Charles Darwin encountered as he tramped along the shores of San Lorenzo Island before returning to Lima.

Archaeologists are inordinately concerned with time. Trying to determine the age of objects and the events and activities they represent is a full-time job for some of us. Until the middle of the twentieth century almost all dating was relative. In other words, all we could say with accuracy was that something was younger or older than something else—a dating method that relied on both stratigraphy and, as already mentioned, association. Stratigraphic layers of deposits—that is, layers of sediments, constructions, artifacts, and other materials—accumulate over time, leaving the earliest materials at the bottom and the most recent at the top. Thus something found at the bottom of a sequence of stratified layers is assumed to be older than something found higher up. All the objects in one layer can be said to be relatively the same age because they are in association with one another. This combination of stratigraphy and association is a powerful way of estimating relative age, but it is not very useful for estimating actual, or absolute, age.

RADIOACTIVITY COMES OF AGE

This all changed in 1947, when Willard F. Libby and his students at the University of Chicago developed a new technique of absolute dating that fundamentally altered the way archaeologists and geological scientists estimate age. Advances in nuclear technology during and after the Second World War included the development of highly sensitive Geiger counters that could measure the radioactivity of various elements. Libby found that the radioactive isotope carbon-14 (^{14}C) decayed at the known rate of 50 percent every 5,730 years (this was known as the isotope's half-life). The other isotopes of carbon, ^{12}C and ^{13}C, are stable and so do not undergo radioactive decay. This observation is important because it means that the ratio of ^{14}C to ^{12}C decreases with the passage of time as the former decays and the later remains constant. All living organisms are made up of molecules comprised primarily of carbon, a minuscule amount of which is ^{14}C. That means that you and I are radioactive and are constantly emitting beta particles as our isotopes of ^{14}C decay. In fact, Libby and his team calculated in 1949 that we emit approximately 12.5 beta particles per minute per gram of carbon.[4] As long as we are alive, though, we keep taking on new molecules of ^{14}C along with all the other carbon we ingest. When we die, however, the stable isotopes of carbon (^{12}C and ^{13}C) stay pretty much constant in our mortal remains, while the radioactive carbon (^{14}C) continues to decay. After 5730±47 years have passed there will only be about half the original amount of ^{14}C remaining in our bodies (assuming there is any part of our bodies still around).[5]

This discovery revolutionized archaeology, because for the first time scientists could say that an ancient object—as long as it was comprised of once-living material—was X years old, plus or minus a certain margin of error. Unfortunately, the method required burning—and thereby destroying—a fairly large quantity of the ancient object so that it could be turned into a gas that could be measured. Scientists worked around this drawback by dating more plentiful, and therefore less valuable, organic material from the same deposit.

So, say we wanted to know the age of the piece of maize that Charles Darwin found on San Lorenzo Island but did not want to destroy the maize itself. We would measure the ^{14}C content of wood charcoal or some other commonly found material from the same deposit and project its age on all other materials in the same general layer that could not be directly dated. Smaller organic objects like a kernel of maize, anything that was either too small or rare or important to destroy, and inorganic objects like pottery or stone (which do not contain ^{14}C and so cannot be dated using this method) all needed to be dated indirectly, in other words, by association.

This is just what archaeologists began to do as soon as there were laboratories and scientists available that could carry out the analyses. The year after Libby and his students discovered radiocarbon dating Paul Mangelsdorf encouraged two archaeology students from Harvard to excavate Bat Cave in New Mexico, where there were known to be large quantities of well-preserved maize cobs. Archaeology graduate student Herbert Dick and botany student C. Earle Smith Jr. began their excavations in the summer of 1948. They initiated a whole new direction in the archaeological study of agriculture in the Americas. Their collections of maize from Bat Cave represented a landmark achievement in the following respects: it was the earliest known and most morphologically primitive maize found to date, it represented the longest sequence (spanning about 3,000 years from the earliest to the latest deposits), and it was the first collection to show an evolution of maize from early to later forms. Mangelsdorf conveyed some of the excitement that the Bat Cave discoveries generated:

> Our analysis of the prehistoric specimens created so much interest among archaeologists that there soon began a flow of archaeological material into the museum which enabled us to study prehistoric maize from many sites throughout the hemisphere. It was directly responsible for me becoming associated with Richard S. MacNeish and working with him on the prehistoric corn from La Perra Cave in eastern Tamaulipas, Mexico, two caves from Infiernillo Canyon in southwestern Tamaulipas, and finally the caves in the Tehuacán Valley in southern Mexico in which prehistoric wild corn was found.[6]

Incredibly, although radiocarbon dating had been introduced several years before, the radiocarbon dates that were obtained from Bat Cave in 1951 were some of the first to directly date actual cobs. Since so many cobs were found in the cave, they could be dated using Libby's method without worrying about losing too much precious material. One sample of cobs (No. 167) dated to 1805±250 years BP and another sample from a deeper layer in the cave (No. 171, mixed with wood) dated to 2316±250 years BP.[7]

Two years later Dick went back to Bat Cave and collected many more maize samples along with dateable charcoal from undisturbed layers. These new charcoal dates ranged from about 5773±209 years BP at the bottom of the deposits to 1658±200 years BP at the top.[8] Although these dates were accepted in the 1950s, many researchers were skeptical, and Mangelsdorf himself pointed out almost twenty-five years later that while the earliest date on the charcoal might well be accurate for the burned wood itself, that one couldn't be certain that the maize associated with the wood charcoal was the same age.[9] Eventually, using various methods of estimating the age of the Bat Cave deposits, Mangelsdorf and others concluded that the earliest maize found there dated to sometime between 4250 and 3450 cal BP, not 6800 to 6310 cal BP, as the associated charcoal date indicated. We will return to Bat Cave when we look at the impact of a new revolution in radiocarbon dating.

In spite of the early dates attributed to the maize found at Bat Cave in New Mexico, most archaeologists and botanists thought that even older maize was to be found farther south, in Mexico. One of the most enthusiastic supporters of this idea was Scotty MacNeish. In 1949 he had excavated samples of maize at the La Perra Cave site in the state of Tamaulipas in northeastern Mexico. MacNeish asked Mangelsdorf to study and report on the maize to see if it might shed some new light on the origins of agriculture. Having just recently looked at the Bat Cave material, Mangelsdorf readily agreed. He examined 87 cobs and many other maize plant parts from La Perra. Willard Libby's lab, which had also dated the Bat Cave materials, did the radiocarbon dating of associated organic material from La Perra Cave and found that maize from the earliest layers dated to 4,578±180 years BP.[10] In calibrated calendar years, this would date to between 5600 and 4950 cal BP—not as old as the earliest Bat Cave dates, but many centuries older than the more generally accepted age estimates for maize from Bat Cave. For MacNeish and Mangelsdorf these relatively early dates were heartening because they showed that maize could be dated to a much older absolute age in Mexico than farther north in the United States. The La Perra maize also showed some morphological similarities (albeit primitive in most cases) with a modern race of maize called Nal-Tel, found in the Yucatan region of southern Mexico.[11]

MacNeish's work at La Perra Cave was just the beginning. He would spend the next fifteen years discovering and excavating many more caves in Mexico and Central America. In 1954 he dug at a series of three cave sites located 120 kilometers south of La Perra—Romero's, Valenzuela's, and Ojo de Agua—all in the Infiernillo Canyon of Tamaulipas.[12] Five years later he had also examined a number of promising locations in Honduras and Guatemala, but none of them proved to have much evidence of early maize. MacNeish finally settled on the Santa Marta Rockshelter—located near the town of Ocozocoautla, Chiapas, in the Grijalva River valley—which he tested with Frederick A. Peterson of the New World Archaeological Foundation. They found very little actual maize there, but what they did find was similar in age to the samples from Tamaulipas, so MacNeish decided to keep looking elsewhere for earlier specimens.[13] In 1960 he began to probe a series of dry caves in the Tehuacán Valley, some 200 kilometers southeast of Mexico City, and it was there that he finally hit the mother lode.

The Tehuacán Valley was an ideal location to continue the search for well-preserved ancient maize because it falls within a rain shadow between two mountain ranges, leading to optimally arid conditions for preserving uncharred plant remains—even better than the dry caves farther north in Tamaulipas. MacNeish tested fifteen caves and found maize remains in five of them: San Marcos, Coxcatlán, Purrón, El Riego, and Tecorral. Over several field seasons in the early 1960s he and his team excavated and screened the deeply stratified layers of accumulated occupation debris in the caves. They recovered tens of thousands of artifacts, animal bones, and plant remains, all of which are carefully described and analyzed in a massive five-volume report published between 1967 and 1972.[14] Maize remains were among the most common type of plant material recovered. To give a sense of the enormity of the collection, here's the list produced by Paul Mangelsdorf: 12,860 whole cobs, 7,819 fragmentary cobs, 46 roots, 506 stalks, 442 leaf sheaths, 282 leaves, 998 husks and husk systems, 12 prophylls, 127 shanks, 384 tassels, 5 midribs, 797 kernels, 83 chewed stalks, and 140 chewed husks. Of the total 24,186 specimens of ancient maize MacNeish's team recovered, 15,000 pieces came from one cave alone: Coxcatlán.[15] No comparable sample has ever been recovered from Mesoamerica, or been so thoroughly studied.[16] What is even more remarkable is that the collection spans such a long time period. Based on radiocarbon estimates at the time, using associated charcoal and other preserved plant parts, the age range of the pieces was thought to span 6,500 years—from about 7,000 years ago to the Spanish Conquest in the 1520s. For MacNeish and Mangelsdorf one of the most important aspects of the Coxcatlán discoveries was the big sample of very tiny maize cobs recovered from the lowest, oldest levels of the cave. One cob in Zone F and 26 cobs in Zone E were about the diameter of a pencil and no more than twice as long as its eraser (about 2 to 2.5 centimeters). Mangelsdorf considered these and the earliest cobs from nearby San Marcos Cave to be his long sought after "wild corn," which he believed was the true ancestor of domesticated maize (figure 4.1).[17]

The thousands of other cobs and maize plant remains at these cave sites gave Mangelsdorf and his students a clear view of the evolution of

FIGURE 4.1. Walton C. Galinat's reconstruction of hypothetical wild maize, fragments of which were thought to have been recovered by Richard MacNeish in early deposits at San Marcos Cave in the Tehuacán Valley, Mexico. (Redrawn by Michael Blake after Galinat's original drawing in Mangelsdorf 1974:180, figure 15.24)

the maize, from its earliest tiny "wild" form to its succeeding larger varieties. Reproduced in almost every archaeology textbook that deals with the origins of New World agriculture is the famous line of five upright maize cobs with a mini "wild" cob from San Marcos Cave on the far left and a large, almost modern-sized cob of Chapalote maize from Coxcatlán Cave on the far right (figure 4.2).[18] This amazing evolutionary snapshot of the Tehuacán Valley maize's family tree is, in most respects, correct even today—but new methods of radiocarbon dating, along with DNA analysis of both maize and teosinte, have changed some of Mangelsdorf's and MacNeish's conclusions—a topic we will examine in chapter 5.

A SECOND RADIOCARBON REVOLUTION

A major breakthrough in radiocarbon dating came with the simultaneous publication of two papers in the November 4, 1977, issue of the journal *Science* announcing the development of a more precise and less destructive method for carrying out radiocarbon dating. One paper by D. Erle Nelson, a nuclear physicist in the Department of Archaeology at Simon Fraser University, and his colleagues described a successful method for directly measuring the ratios of the three carbon isotopes in an ancient wood sample.[19] The other paper, by nuclear physicists

FIGURE 4.2. Drawing reconstructing one of Paul Mangelsdorf's famous photographs depicting the evolution of corn using selected specimens from Richard MacNeish's excavations at San Marcos and Coxcatlán Caves in the Tehuacán Valley. The smallest specimen on the left was though to be wild maize. The largest specimen on the right represents nearly modern maize, measuring approximately 13 centimeters in length. (Redrawn by Michael Blake based on the photo in Mangelsdorf 1974:182, figure 15.25)

C. L. Bennett, Harry E. Grove, and their colleagues, further explained how, by using an accelerator beam, they could strip away the unwanted nitrogen (^{14}N) atoms and directly measure the relative amounts of ^{14}C, ^{13}C, and ^{12}C atoms using a mass spectrometer—hence the name AMS, which stands for "accelerator mass spectrometry." In fact, they found they could count as few as three atoms of the exceedingly rare radioactive carbon ^{14}C even when they were embedded in 10^{16} atoms of ^{12}C (that's a ten followed by sixteen zeros!).[20] Within a few years the technology was being developed to carry out AMS radiocarbon dating at a number of labs around the world.

One huge advantage of this type of dating is that very small samples could be dated: 5–100 milligrams compared to the 30–200 grams that had to be used in the conventional beta particle counting method developed years earlier by Willard Libby. This is especially important for small objects, such as ancient plant remains. Prior to AMS dating, it was necessary to incinerate whole maize cobs in order to get enough carbon to measure. Archaeologists and botanists were reluctant to do this because the samples were usually too precious. Conventional radiocarbon dating, for example, would have consumed all of the earliest tiny cobs from Coxcatlán Cave just to get one date. Therefore, prior to the 1980s most archaeological maize specimens could only be dated by association—that is, chunks of charcoal or other materials found close by the ancient maize were dated and those associated dates were used as proxies for the age of the maize samples.

One of the earliest applications of AMS to date maize was carried out by a fellow graduate student of mine, Wirt "Chip" Wills, for his doctoral dissertation project at the University of Michigan. Chip re-excavated a portion of the Bat Cave site in the early 1980s and recovered new samples of maize from the intact cave deposits. He immediately had some of them dated using the new AMS method. The earliest direct date that he obtained on maize from one of the lowermost layers was 3010±150 BP. This translates to a calendrical date with a median value of 3185 cal BP and an upper to lower range of 2795 to 3555 cal BP, with a probability level of 95.4 percent (figure 4.3).[21] The earlier end of this range is not too far off the later end of the 4200–3450 cal BP age range that Mangelsdorf had estimated after rejecting the even earlier dates on associated charcoal, which dated back as far as 6800–6310 cal BP.

A new dating controversy emerged in 1989, when another team decided to directly date maize samples from some of the earliest maize-bearing deposits in the Tehuacán Valley. Austin Long from the University of Arizona radiocarbon lab, along with the botanist-archaeologist Bruce Benz and their colleagues, obtained permission to date a sample of maize cobs from Coxcatlán and San Marcos Caves and found something truly startling. They discovered that the maize cobs that had been assigned to the Coxcatlán Phase (spanning the period from 7000–5350 BP) actually turned out to be, at the earliest, only 4700±110 BP.[22] This showed that the maize was very likely not as old as had been previously thought—perhaps by as much as 2000 years. Just as with Bat Cave, older charcoal had likely been mixed with younger maize—and direct AMS dating of the actual maize samples themselves was able to sort it out.

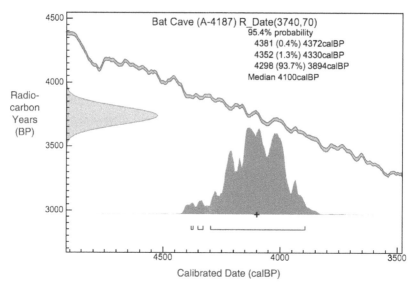

Radio-
carbon
Years
(BP)

Bat Cave (A-4187) R_Date(3740,70)
95.4% probability
4381 (0.4%) 4372calBP
4352 (1.3%) 4330calBP
4298 (93.7%) 3894calBP
Median 4100calBP

Calibrated Date (calBP)

FIGURE 4.3. An example of a printout from the OxCal website, showing the radiocarbon calibration curve estimate for a maize AMS date sample from Bat Cave, located on the San Augustin plain in New Mexico. This sample (A-4187) was originally dated and calibrated to the tree ring sequence by Wirt Wills (1988). A recent OxCal calibration shows that the estimated mean radiocarbon age of 3740±70 years BP is about 360 years younger than the true age estimate, which gives a median age of 4100 cal BP. There is a 95.4 percent probability that the true age of this sample falls somewhere between 4382 and 3894 cal BP. (Calibration graph produced by OxCal V4.2.4 using IntCal13, C. Bronk Ramsey 2014)

Many archaeologists have worried that dry cave sites, where maize and other plant remains are often well preserved, make great homes for burrowing rodents, insects, and other animals, precisely because they are so dry (see chapter 6). Unfortunately for archaeologists, these cave denizens can, over time, mix large quantities of earth, and the objects it contains, as they burrow beneath the cave floor. The human occupants of the caves often dug storage pits, hearths, and other structures in the floors of their homes. These disturbances could open up holes leading from the surface down to deeper layers, increasing the likelihood that younger objects—including plant remains, such as maize cobs—originating in the topmost layers might work their way down toward lower, earlier strata and eventually get mixed in with earlier materials. It is even possible that older materials could be carried upward, toward the surface, although this is less likely.[23]

This set of findings gave pause to claims of very early domestication of maize and led some archaeologists to think that maize agriculture may not have the antiquity that had been previously claimed.[24] Many new studies aimed at using AMS radiocarbon dating to redate ancient maize samples from sites in the United States, Mexico, and points farther south—samples that had previously only been dated indirectly by association—have helped to refine the overall picture of early maize agriculture. To date, over six hundred samples of maize, both macroremains (large parts of the plant, such as kernels and cobs) and microremains (almost invisible parts, including pollen and starch grains and phytoliths), have been directly dated using the AMS technique, and more are being added to the roster every day. One region where AMS dating is especially prevalent is the southwestern United States. More than half of all AMS-dated maize comes from the United States, and about 80 percent of these US samples come from Arizona, New Mexico, Utah, and Colorado.

This wealth of information provides a stark contrast with the data available for other parts of the Americas. As we move farther south from the US-Mexico border region the number of AMS-dated maize samples decreases sharply. In fact, all of the countries in Latin American combined—including Mexico and Honduras—have just over 180—fewer than we have for the four aforementioned states in the US Southwest alone. This is sure to change as the AMS method becomes less expensive and more widely used and as archaeologists realize the increased reliability of direct dating versus dating by association.[25]

One result of this discrepancy is that we know a great deal more about the dispersal of maize in central and northern Mexico and the United States than in southern Mexico, Central America, and South America. A priority for archaeologists working in Latin America is to correct this problem. During the past ten years or so some researchers have been taking on the work of AMS-dating ancient maize samples, and this continues to add to our knowledge of maize's early history. Bruce Smith, Bruce Benz, and their respective colleagues have helped by redating samples from MacNeish's collections from the 1950s and 1960s as well as dating those that haven't been tested yet, and Dolores Piperno and Kent Flannery have AMS dated two of the tiny cobs from Guilá Naquitz.[26] South of Mexico there are fewer such efforts, but this is changing as older collections housed in museums are being reanalyzed and dated. Archaeologists working in Peru, for example, have access to older collections and are also excavating new maize samples from some of the driest open-air sites in the Americas, which have plant remains that have been better preserved than anywhere else in the world.[27] Dating of maize and many other domesticated plants by association will be increasingly replaced by AMS direct dating, and this promises to contribute new answers to the questions of how and when agriculture and domestication developed and how and when maize was dispersed into North and South America.

MAPPING THE AGE OF MAIZE'S SPREAD

In 2006, in an effort to answer some of these questions, I mapped the distribution of the earliest-known, AMS-dated archaeological maize samples in order to see if there were any clear patterns of maize distribution based on the data that was currently available at the time.[28] The patterns were clear, yet woefully incomplete, because there were still so many regions of the Americas that had so few or no directly dated samples. Since then, many new dates have become available—including those mentioned above and more than ninety dates that Bruce Benz and I have been able to gather since 2008 (see note 25)—and they have been added to the mix, as shown in map 4.1.

The distribution of the earliest directly dated maize remains throughout the Americas demonstrates two things. First, it shows that the oldest macroremains of maize are found very close to the original natural habitat of the teosinte subspecies *Zea mays* ssp. *parviglumis* and *mexicana* that gave rise to maize. Second, it shows that there is a time gradient as domesticated maize spreads north and south from its original homeland.

As I've mentioned, the earliest securely dated maize macroremains were discovered at Guilá Naquitz Cave in Oaxaca and dated to circa 6230 cal BP. These tiny cobs were similar to teosinte in many respects, and we will examine their morphological characteristics in chapter 5. Since it is likely that the earliest maize did not originate in Oaxaca, but instead was domesticated a few hundred kilometers to the northwest in the Balsas River region, there could well be earlier maize cobs—which could look even more like teosinte—in some as-yet-undiscovered cave or rockshelter site. Even so, as we look at the distribution of early maize ages on the map we can see that the next earliest macroremains come from the Tehuacán Valley caves and, after that, the Tamaulipas caves. The three oldest sets of maize remains all come from dry cave sites and indicate movement to the south and north of the Balsas River region. As

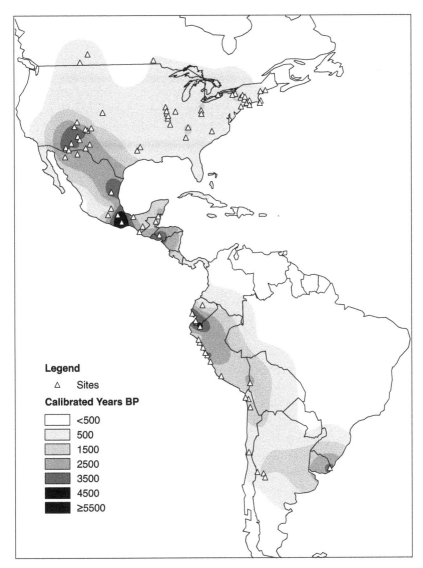

MAP 4.1. The age distribution of archaeological sites with directly dated maize macro- and microremains. The 1000-year-intervals are based on the oldest date at a site. All sites located within a 50-kilometer radius of the oldest dated site in a region are excluded. (Created by Michael Blake and Nick Waber using sources referenced in Ancient Maize Map, http://en.ancientmaize.com/)

we will see later, this may be more a reflection of the locations of dry cave sites than an indication of where and when maize was first transported and grown.

Heading north beyond these three dry cave locations, our map shows the next earliest maize macroremains occurring in Arizona and New Mexico. In the past few years archaeologists have discovered and excavated a series of open-air sites in these states, which represent the seasonal occupations of Preceramic peoples. The sites show that these peoples must have been growing maize for at least part of the year because researchers have found a surprisingly high frequency of charred maize fragments. The earliest maize found at these sites is only a century and a half younger than the maize from the Tamaulipas caves in Mexico. One of these sites, the Old Corn Site, located near the Arizona border in central New Mexico, has the earliest-known maize samples in the United States: circa 4205 cal BP.

Located on the margins of the Santa Cruz River as it flows through the western suburbs of Tucson, Arizona, just 300 kilometers to the southwest of Old Corn, is a series of ancient village sites with shallow housepit depressions and refuse deposits, many of which have yielded abundant maize remains. The two sites with the earliest samples so far—Clearwater and Las Capas—have ancient maize directly dating to circa 4060 cal BP and circa 4005 cal BP, respectively. These are only a few of many sites in these two states that have maize samples that are directly dated between 3,000 and 4,100 years ago, the earliest of which are from charred maize remains recovered by flotation sampling of soil from house and campsite refuse deposits in open-air sites—not caves. This confirms that maize was present in the US Southwest even earlier than the evidence from cave sites in the same region, such as Bat Cave, would have suggested.[29]

This has several implications for the discovery of early maize in the regions to the south of teosinte's homeland. Since maize was on the move to the north, reaching the US Southwest by about 4,000 years ago, we might expect it to have also moved southward by the same time, if not even earlier. The earliest AMS-dated maize macroremains in Mexico, south of Guilá Naquitz in Oaxaca, come from several Early Formative period village sites on the Pacific Coastal Plain in southern Chiapas, next door to Guatemala, which John Clark and I excavated in 1985 and 1990 during the first two seasons of our Mazatán Early Formative Project.[30] The oldest maize sample that we dated was from the site of San Carlos and it came in at 3605 cal BP. Three neighboring

FIGURE 4.4. View of work in progress at El Gigante Cave in the western highlands of Honduras. Maize macroremains excavated at this site by Tim Scheffler in 2000–2001 as part of his doctoral research are the oldest and best preserved in Central America. (Photograph courtesy of Ken Hirth)

sites—Paso de la Amada, Chilo, and Aquiles Serdán—also had maize kernels that were more than 3,000 years old. Since these samples are more than 2,600 years younger than the Guilá Naquitz cobs, it is very likely that maize had made its way to Chiapas, neighboring Guatemala, and points farther south much earlier than the preserved remains in these few sites would suggest. We can't assume that these samples represent the earliest occurrence of maize in southern Mexico—they are simply the earliest macroremains we have dated so far. As we shall see, there is an accumulating body of evidence suggesting that maize moved southward into Chiapas and beyond into Central America much earlier.

Located in the uplands of west-central Honduras, approximately midway between the Pacific and Caribbean coasts, the dry rockshelter site of El Gigante has produced the largest and best-preserved samples of dry plant material, including maize, found anywhere in Central America (figure 4.4). The site was excavated by Tim Scheffler in 2000–2001 for his doctoral project at Penn State University, under the supervision of archaeologists Ken Hirth and David Webster and archaeobotanist Lee Newsom. Scheffler discovered deep deposits in the rockshelter, layer upon layer of refuse that had accumulated during ten millennia of occupation, beginning about 12,000 years ago. Maize cobs were plentiful throughout the Archaic period and later deposits—but of the two that were dated for Scheffler's project, the oldest yielded an age of only 2260 cal BP.[31] Hirth and Webster suspected that some of the maize cobs from the deeper deposits were much older, but no additional samples had been dated during the decade following the excavations.

Bruce Benz and I thought we might be able to help, so as part of our maize-dating research project we contacted the Penn State team to suggest running more AMS dates on maize cobs from the site as part of our research grant. Fortunately, they readily agreed. We traveled to Penn State, where the collection was stored, and selected forty-nine maize cobs for dating—choosing samples that were distributed from the bottommost layers to near the surface of the rockshelter's floor. The youngest we dated was about 510 cal BP, while the oldest was about 4370 cal BP. In fact, seven of the cobs from the deepest layers came in at over 4,000 years old—about 1,000 years earlier than the maize we dated in the early village sites in Chiapas. Again, these dates are surprisingly similar to the earliest dates from Arizona and New Mexico in the north.[32] But, as old as these El Gigante maize cobs are, they don't represent the earliest directly dated maize remains so far discovered in Central and South America.

Apart from directly dated maize macroremains, there are two additional types of evidence for earlier maize south of Mexico. One comprises radiocarbon dates on charcoal and other organic materials associated with maize macroremains (i.e., indirect dating). The other comes from radiocarbon dates on organic material associated with microremains (some of which can be considered direct dating, but most of which is indirect dating by association). Let's look first at the associational dates for maize macroremains before turning to the microremains.

There are very few associational dates for the southern regions that are earlier than the AMS dates from El Gigante Rockshelter. One of them comes from the Arenal Reservoir in Costa Rica. This sample consists of a maize kernel that was dated by association to 5090 cal BP.[33] This is only about eight centuries earlier than the earliest direct dates in Arizona and New Mexico, which is not unreasonable, considering that Costa Rica is only about 300 kilometers farther to the south of the Balsas River region than southern Arizona is to the north (about 2,000 kilometers and 1,700 kilometers, respectively) and that Costa Rica's biogeography is much more similar to that of maize's homeland than is the biogeography of the US Southwest. Therefore, because early maize could have adapted more easily to southerly environments, its southward spread into Central America is likely to have been faster and earlier than its spread into North America. Still, because the Arenal sample is indirectly dated, we can't be sure of its exact age. It could be younger . . . or older.

The only earlier dates by association come from the work of Duccio Bonavia and Alexander Grobman in Peru, and two of these dates are worth mentioning. One comes from the site of Cerro El Calvario (PV32–1) in the Casma Valley along the north-central coast of Peru. Maize was found there in association with non-maize plant remains that were dated to 6875 cal BP, six centuries older than the Guilá Naquitz cobs in Mexico.[34] It would be relatively easy to confirm the age by directly dating the maize from Cerro El Calvario using AMS dating, and hopefully this is something that can be done soon, because without it the probability that the sample is misdated is simply too great.

The other early date comes from the Late Archaic period (Preceramic) site of Los Gavilanes in the Río Huarmey Valley, also located on the north-central coast of Peru. At this site, first sampled by David Kelley in the 1950s and later excavated by Duccio Bonavia between 1967 and 1979, large quantities of maize were recovered in and around what appear to be storage pits.[35] Originally, the excavators dated organic material (not maize) from one of these pits and obtained a date of 4610 cal BP.[36] This date is somewhat more likely than the very early one from Cerro El Calvario, at least in part because it would mean the maize is younger than the earliest-known directly dated maize from Mexico. It is not too hard to conceive of maize moving all the way into South America within 1,600 years of its appearance in Guilá Naquitz. However, as with the samples from the Casma Valley, it is still preferable to do direct AMS dating on samples where possible, rather than to rely on dates by association.

Reasoning that it would be a relatively easy matter to directly date some of the maize remains from Los Gavilanes, Bruce Benz and I decided to track down some of the samples from the original excavations carried out at the site in the 1950s under the direction of David Kelley. David Kelley and Jane Kelley, Harvard alumni who were then professors emeriti in the Department of Archaeology at the University of Calgary, kindly gave us permission to examine the maize samples, and they were excited to hear about our plans to carry out AMS dating on a few of them. Benz and I traveled to the Harvard University Herbaria, where the specimens were stored, and, under the watchful eye of Emily Wood, the Senior Collections Associate, we were able to collect several small samples for AMS dating.[37] Eight of these were submitted to the NSF-Arizona AMS Lab for dating, and all returned dates were much younger than expected. The oldest was only 2230 cal BP, and the remaining seven averaged out to about 650 cal BP. It seemed to us that, although a major part of the occupation at Los Gavilanes may actually have been during the Late Archaic period—in other words, older than 4,000 years—most of the maize recovered from the site could well be from later peoples who used the location to process and store their maize.

One of the most promising techniques for studying the spread of maize is dating three types of microscopic particles produced by most species of plants: pollen, phytoliths, and starch grains. We will look at the kinds of methods and techniques of analysis that researchers have developed to study plant microremains in chapter 6. In the meantime I want to introduce the topic of how these microscopic remains have contributed to our understanding of the timing of maize's spread throughout the southern regions of the Americas. All three of the aforementioned types of microremains are produced in abundance by a wide range of plants and have been successfully used to study ancient environments, agriculture, and subsistence. Their dating is most often done by association, in a similar fashion to the method used for dating macroremains. All of these microremains can be identified by their shapes, sizes, and, in some cases, light-scattering characteristics, and they can also be categorized by family, genera, and sometimes species. Using phytoliths, researchers are even able to identify distinctive plant parts—such as cobs or leaves in the case of maize—because different cells produce distinctly different phytolith shapes.

Some of the microremain evidence that researchers can draw on to understand the history of maize dispersal throughout the Americas comes from environmental deposits, such as lake-bed or swamp sediments, rather than from archaeological sites where people actually lived. When pollen is dispersed naturally and falls into bodies of water—such as lakes, streams, and lagoons—some of it eventually sinks to the bottom and becomes part of the accumulated sediment record. If sufficient quantities of other organic matter also accumulate in the sediments, then that material can be directly dated—and now, with AMS dating requiring very small samples, it is relatively easy to find enough organic material to reliably date. However, because pollen grains are so minute, it is often difficult to get enough pollen from a given layer to date directly, so most researchers rely on the association between pollen and other dateable organic material (e.g., peat, charcoal, twigs, and leaves) found in the same sample to obtain a date. Assuming the pollen is in fact contemporary with the material being dated, and not intrusive from either earlier or later deposits, the dating procedure can give an accurate estimate of the pollen's age.[38]

The same is true for phytoliths, which are small silica particles that form within plant cells, helping to give the plant its rigidity and, in some cases, its hardness. Since phytoliths are not designed to be dispersed in the wind, like many types of pollen, they make their way into the soil and lake sediment deposits when the original plant material decays. Whereas pollen might blow into a lake from a maize plant growing in a nearby field, phytoliths are more likely to be washed in with plant parts after rains or flooding, or fall into the water as part of the soil erosion process. They might also blow in as wind picks up soil from an exposed field that already contains phytoliths. If the fields are burned, then some of the phytoliths might exhibit signs of fire damage.

Both pollen and phytoliths have been recovered from lake sediment cores in Central and South America, and, in the absence of maize macroremains, such as kernels or cob fragments, microremain research has extended the estimates of maize's antiquity in these regions. Dozens of studies have been carried out since the 1960s, and with more accurate AMS dating in recent years we have increasingly reliable estimates of the ages of pollen and phytoliths associated with dateable materials recovered from lake and swamp sediment cores.

One of the earliest *Zea* examples in southern Mesoamerica comes from the San Andrés site, located near an ancient lagoon just 5 kilometers northeast of the ancient Olmec ceremonial site of La Venta in Tabasco, Mexico.[39] The archaeologists excavated parts of the site in the traditional manner, using trowels and sieves, but they also took deep sediment samples by driving—or, more accurately, vibrating—long, 8-centimeter diameter aluminum tubes into the waterlogged sediments. Using this technique, called "vibracoring," they were able to recover small wood samples from one core column that reached a depth of just over a meter below the modern ground surface. The samples produced an AMS date of 7100 cal BP. Associated with this wood, in the same layers of the core sediments, were small pollen grains identified as members of the Poaceae family of tropical grasses. Because they were relatively large compared with the average pollen grain size for most species of tropical grasses, the grains were most likely produced by either teosinte or early maize. Since the Gulf coast of Tabasco, where San Andrés is located, is outside the natural habitat for teosinte, this pollen probably belongs to an early form of domesticated maize. More recently the same archaeologists recovered a small number of phytoliths that they have identified as maize from the same position in the core samples. These two lines of evidence suggest that the practice of maize cultivation had reached the Gulf coast lowlands some 900 years before maize appeared in the form of the tiny, teosinte-like ears at Guilá Naquitz in Oaxaca Valley.[40]

Maize pollen has been identified in many lake sediment cores taken in Mexico, Belize, Guatemala, El Salvador, Honduras, Costa Rica, Panama, Colombia, and Ecuador, all dating between about 3500 and 6500 cal BP. But an even earlier sample has been identified at a cave site named Cueva de los Ladrones (Cave of the Thieves), located about 25 kilometers inland from Parita Bay, along the Pacific coast of Panama. Beginning 8,000 years ago, the cave was regularly used by Archaic period peoples who practiced hunting, gathering, and agriculture. Archaeologists Anthony Ranere, Dolores Piperno, and Richard Cooke recovered several grinding stones from the earliest layers in the cave. Careful washing of these stone tools allowed the researchers to extract microscopic sediments that had adhered to the grinding surfaces. The initial analyses of these sediments revealed that they contained both maize pollen and phytoliths, and later work on tools from the site recovered microscopic samples of maize starch grains as well. Because the stone tools on which these microremains were found were recovered in close association with charcoal samples that were radiocarbon dated to circa 7800 cal BP, it can be inferred that maize was being used at the southern tip of the North American continent by that time.[41]

Dolores Piperno and her colleagues have recently reported that grinding stones from the Aguadulce Rockshelter in Panama harbored evidence of starches from maize, along with manioc (*Manihot esculenta*), yams (*Dioscorea* sp.), and arrowroot (*Maranta arundinacea*), all from a stratigraphic layer dated as early as 7750 cal BP.[42] On another stone tool only slightly younger they recovered starch grains from domesticated chili peppers.[43] For dating enthusiasts, one of the novel aspects of the Aguadulce Rockshelter dating effort was that the researchers used AMS dating of phytoliths to obtain age estimates for each stratum. (The age of the phytoliths is determined by combining thousands of individual phytoliths from different plant taxa. This technique has not been used very often, but it does hold the potential for directly dating phytoliths of a known species from a single large deposit in order calculate a more precise age of a plant based on direct dating.)[44]

These discoveries, and many more that we will explore in chapter 6, suggest that maize was poised to move (or rather, be moved) into South America much earlier than the directly dated macroremains would suggest. As in Central America, several radiocarbon measurements of charcoal associated with pollen in lake sediment cores from northern South America indicate early movement of maize into the region. The earliest of these cores comes from a small lake near Hacienda el Dorado in Colombia, where maize pollen is associated with a radiocarbon date of 5360 cal BP—substantially more recent than the earliest maize microremains from Panama and almost 900 years younger than the earliest maize cobs from Oaxaca.⁴⁵

Farther to the south, in Ecuador and Peru, phytoliths and starch grains have been found and dated to more than 5,000 years old as well. Led by Karen Stothert, archaeologists working at the Vegas site, located only 4 kilometers from the Pacific coast in Ecuador, excavated a site that represents one of the earliest coastal villages in the Americas. There, Stothert and her team discovered evidence of residential remains, including house features, tools, food refuse, and human burials, dating between 11,000 and 7500 cal BP—a period during the Early Archaic, called the Las Vegas phase. Phytoliths of many plant species were common in the soils of the site, and distinctive maize phytoliths were associated with layers that were radiocarbon dated to circa 7960 cal BP, toward the end of the late Las Vegas phase.⁴⁶

The Las Vegas site and its neighbors in coastal Ecuador predate, by as much as 1,500 years, a sequence of occupations that, since the 1950s, have been labeled the Valdivia culture—early villages that existed from about 6000 to 3700 cal BP. Taken together, the Vegas and Valdivia peoples represent some of the earliest agriculturalists in either North or South America who are known to have been sedentary and to have relied, to a certain extent, on a range of cultivated plants. Although no maize macroremains have yet been discovered at the open-air Valdivia sites, archaeobotanists have found the phytoliths of maize and many other domesticates—in greater amounts than at the older Vegas phase sites. The ages of maize phytoliths from the Valdivia phase sites were determined by dating associated charcoal fragments, marine shells, human bone, and the phytoliths themselves—similar to the procedures used for dating the samples from the Vegas-phase sites. Although the possibility exists that some younger phytoliths somehow worked their way downward into the earlier deposits, the archaeologists who excavated the sites and studied the plant microremains don't think this very likely, first, because phytoliths are not thought to move around very much in the soil, and second, because most of the samples that were studied come from deposits that appeared to have been undisturbed.[47]

More recent work on the other side of the Andes Mountains in Ecuador has recovered maize microremains that date to about 5,000 years old. Francisco Valdez and his team have been excavating the impressive site of Santa Ana-La Florida for the past ten years, and they have made some remarkable discoveries. Sitting alongside one of the tributaries of the Marañon River, which in turn flows into the Amazon, Santa Ana-La Florida is contemporary with the coastal Valdivia sites but has its own ceramic traditions. Ornate stirrup-spout vessels, stone bowls, and other superbly made artifacts have been recovered from elaborate sealed tombs near the center of this ceremonial site. As part of her PhD research at the University of Calgary, Sonia Zarrillo identified maize starch grains and phytoliths from residues taken from the surfaces of several of the mortuary offerings found at the site. The tomb is dated to at least 3980 cal BP, and because the artifacts were sealed inside, the age of the maize microremains is secure. However, there is even earlier evidence for the use of maize at the site. A piece of pottery recovered from a trash deposit near the site's center had charred residue adhering to it—and there was enough of it to both look for microremains and obtain an AMS radiocarbon date. The carbonized organic material dated to 4980 cal BP, and embedded within these residues were at least sixteen maize starch grains (figure 4.5).[48] This means that the date of the organic material also counts as a direct date for the maize, since there is no chance that these starch grains could have penetrated the carbonized residues after deposition.

Farther south, in southern Peru, maize phytoliths and starch grains were recovered by Linda Perry and her team at the Waynuna site, an ancient high-elevation village over 100 kilometers inland from the coast. The site sits at 3,625 meters above sea level and is well out of the range of the coastal lowlands, where most early maize phytolith and starch grain samples have been reported so far. Many of the maize microremains found at Waynuna are associated with a Preceramic period house—but the earliest samples, which date to 4030 cal BP, come from just below the house floor.[49] The presence of maize microremains in clear association with a Preceramic house floor that is more than 4000 years old and located at such a high elevation suggests that maize agriculture must have been present in the Andean highlands even earlier. It is unlikely that the maize could have been successfully grown at Waynuna unless it

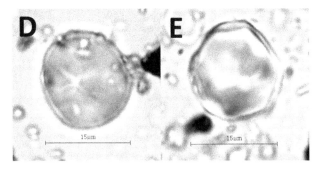

FIGURE 4.5. Maize starch grains embedded within carbonized organic material adhering to a piece of broken pottery recovered by Francisco Valdez from a trash midden at the site of Santa Ana-La Florida in Ecuador. The carbonized material was directly dated using AMS dating and turned out to be about 4980 cal BP plus or minus one to two centuries (Zarrillo 2012:213–214). (Photograph courtesy of Sonia Zarrillo)

already had a long history of adaptation to such high elevations, which, even today, constitute the extreme limits of maize's tolerance.[50]

Dating the domestication and spread of early agriculture is not always straightforward, and this has led to a number of contentious debates over the years. We have seen that the direct AMS dating of the earliest maize macroremains changed the perception of when maize might have initially appeared at the Tehuacán Valley sites. We also know that maize macroremains are not often preserved in the humid tropics, so new methods have been developed to try to determine the age of maize agriculture in places with these climates. The dating of the earliest maize in South America has sparked further debate. Some have argued that maize phytoliths and pollen in natural contexts (such as lake sediment deposits) and cultural deposits (such as village sites) provide sufficient evidence for maize cultivation. Both pollen grains and phytolith particles are sufficiently robust to preserve over very long spans of time (tens of thousands of years) and can be considered to be safely stable in their contexts of deposition (that is, they don't move around too much after being deposited). Others have argued that in the absence of other lines of evidence (such as the macroremains of the domesticated plants) we should be more cautious about assigning age determinations based on microremains alone.

Both sides of this debate have a point. On the one hand, we do know that some ancient deposits are intact, and there is little likelihood that older or younger materials moved around much. On the other hand, it is clear in some cases that there was disturbance to deposits over time, and we can't always see the visual evidence of this. The actions of insects, rodents, and even humans can churn up deposits in unexpected ways, reversing stratigraphy and mixing up the original associations of materials. Trees often have massive root systems that can disturb deposits and lead to the movement and mixing of soils—especially if a tree burns and the fire penetrates the root systems far below ground. By excavating carefully, archaeologists are often able to observe the evidence of such mixing and disturbance, which they can then take into consideration when collecting samples for dating. Both microremains and macroremains can be subjected to disturbance, and we shouldn't expect one or the other to be more susceptible.

When dating ancient objects, including plants, most researchers agree that it is preferable to directly date the object under study. Therefore Piperno and Stothert's direct AMS dating of an amalgamated sample of hundreds of *Cucurbita* phytoliths is the next best thing to dating an actual fragment—a seed, chunk of rind, or stem—of ancient squash. The same would be true for any other plant, including maize.

The relatively recent method of directly AMS dating ancient plant microremains trapped in charred residues found on pottery or other artifacts—work that Sonia Zarrillo was doing at Santa Ana-La Florida in Ecuador—has great promise. But it means that archaeologists will, in the future, have to be much less fastidious about washing and scrubbing the earth-covered artifacts they find in tombs, houses, and trash deposits if they hope to use this kind of "microdeposit" to directly date the minute remnants of the plants ancient peoples used.

Another successful example of this approach also comes from Ecuador. John Staller used this method on pieces of pottery excavated at the Valdivia period site of La Emerenciana, located on the south coast of Ecuador. Some of the interior surfaces of the pottery sherds had a layer of crusted carbon, which Staller and his colleague Robert Thompson thought they might be able to search for identifiable phytoliths and then directly date using AMS. They scraped some of the carbon from the rims of several sherds and submitted two samples for radiocarbon dating, while samples of the remaining carbon were processed to look for phytoliths. Two of the samples contained distinctive maize phytoliths, and the earliest yielded an AMS date of 4200 cal BP. Because the phytoliths were extracted from the residue of maize that had somehow been burned directly onto the rim of a ceramic vessel (perhaps an early case of too many cooks?), the unassailable association of the maize phytoliths, the distinctive late Valdivia period ceramic vessel, and the ^{14}C age estimate leaves no doubt that the phytoliths are as old as they claim. Even if the ceramic vessel had been found in mixed deposits, this still would be true.[51]

However, Staller and Thompson went on to make another inference that has generated some contentious debate. They suggested that the La Emerenciana maize phytoliths actually represent the earliest instance of maize in Ecuador and that early microremains associated with pottery vessels is a much more reliable way of identifying early maize use than dating phytolith assemblages from soils. Using the date from the pottery vessel mentioned above, they contend that maize may have been introduced to Ecuador as recently as 4,500 years ago rather than circa 8,000 years ago, which is suggested by dates associated with the abundant maize phytoliths recovered from the Vegas site and other Valdivia period sites farther to the north.[52]

One of the people arguing that maize was introduced earlier into South America, and Ecuador in particular, is Deborah Pearsall. She and her colleagues examined a collection of milling stones excavated from an early Valdivia house floor at the site of Real Alto in Ecuador to determine if microremains could be recovered from artifacts in a securely dated house. The house is Structure 20, and it has been dated by pottery association—that is, the ceramics found in the house deposits are exactly the types and styles known to date to between 4750 and 4350 cal BP.[53] Unfortunately there was not enough charcoal or other organic material in the house that could be directly dated, so they had to rely on the time-honored technique of dating by association, which Darwin used when he found maize remains mixed with shells, pottery, and other artifacts on the coast of Peru 180 years earlier.

In the following chapters we will look more closely at both types of ancient maize remains—macro and micro—and examine some of the new methods of analysis that researchers are bringing to bear on their quest to trace the origins and spread of ancient maize agriculture.

CHAPTER 5

Maize through a Magnifying Glass

Macroremains

For many generations archaeologists have recovered plant remains from the buried refuse layers in ancient settlements. But it is only relatively recently that more attention has been paid to the systematic recovery and identification of these remains. The archaeological specialists that devote their time and energy to this endeavor are known as paleoethno-botanists, and they have revolutionized the collection and study of ancient plant macroremains—plant remains that are large enough to be seen with the naked eye—with a recovery technique known as "flota-tion."[1] There are many different methods of flotation, but the basic principle is the same for all of them: soil from an archaeological deposit is placed into a large container, or drum, that is filled with water and agitated. The plant remains—which are usually charred—are somewhat buoyant and rise to the surface, where they can be skimmed off. Mean-while, the fine soil, now turned to mud, passes through a submerged screen, which captures any artifacts, as well as the plant remains that didn't float. Deborah Pearsall's excellent history of this technique tells how it became increasingly well known in the late 1960s and the early 1970s after the publication of two important articles, one by Stuart Struever in 1968 and another by David French in 1971.[2]

I mentioned in the introduction that I was first introduced to this mar-velous technique in the summer of 1976 as a novice field-crew member participating in the Mimbres archaeological project in New Mexico. The more experienced crew members occasionally enlisted my help with the hard labor of doing flotation analysis, that is, dumping large bags of soil collected from the day's excavations into an old 55-gallon drum filled with water. Suspended inside the drum was a washtub fitted with a fine mesh screen for its bottom. We poured the soil into the tub and then gently agitated it until the soil washed through the screen, leaving the larger objects behind. This flotation machine, newly designed by Paul Minnis and Steve LeBlanc, worked wonderfully well in the arid conditions of the Mimbres Valley.[3] Charred plant remains would, in theory, float to the surface, where they could be collected with a fine cheesecloth strainer. The first time I tried this a few charcoal fragments floated to the surface (this is called the "light fraction") but not much that was recognizable. All the contents gathered from the bottom of the washtub (the "heavy fraction"), along with the carefully wrapped cheesecloth with its precious charcoal, were dumped onto a piece of newspaper and set on shelves in the lab to slowly dry. After a few days, one of the project's two graduate student paleoethnobotanists, Paul Minnis and Christine Hastorf, would take down one of the dried sample containers and patiently sort through it, separating the identifiable fragments of charred plants—most of which were small chunks of burned wood produced in ancient kitchen hearths— from the tiny pieces of household debris—which included animal bones, stone flakes, potsherds, and miscellaneous rock fragments. One evening during a regular session of lab work, Christine called me over to see some charred fragments she had sorted, and there among the pieces of firewood charcoal and roots were tiny blackened kernels—shriveled and cracked, but clearly recognizable as maize. The paleoethnobotany team occasionally discovered charred pieces of maize cobs among the debris and many details about maize and other plants grown and eaten by the ancient Mimbreños could be inferred from these fragmentary remains.[4]

This method of recovering ancient plant remains from archaeological sites is now used around the world, and it consistently produces some of the best sources of information we have about past plant use. It is particularly useful in contexts where economically important plants were likely to have been only partially burned, or charred, so that they retained much of their original shape and characteristics, but not so badly burned that they turned to ash and disappeared. Charring is fortuitous in that it drives off the moisture in the plant and destroys many of the organic nutrients that would make the material attractive to microbes and fungi that would otherwise simply eat it. In charred form, macroremains can remain in the soil indefinitely, withstanding the ravages of the organic decay process.

Besides charring, three other natural preservation processes have led to the discovery of ancient plant remains in archaeological contexts:

(1) desiccation in dry caves and deserts,

(2) waterlogging in swamps, lakes, streams, bogs, and mud-slides, and

(3) freezing in snow and ice.

The first of these has proven to be the most important in the recovery of early maize macroremains throughout the Americas. As discussed in chapter 4, dry cave sites, and especially caves in desert environments, have provided the largest and most complete samples of ancient plant remains so far recovered. As with charring, in super-dry environments moisture is driven off the organic material, eventually leading to complete desiccation. If there is no moisture in and around the cells of the plant material, microscopic organisms cannot survive and reproduce in sufficient numbers to consume it. Cave sites where large quantities of well-preserved maize and other plant remains have been discovered are found in the United States, Mexico, Honduras, Peru, and Chile. At some cave sites the preservation is so good that archaeologists remark that they have a hard time believing that the plant parts they are collecting are indeed ancient. In some cases, maize cobs and kernels that are thousands of years old, as well as the remains of many other plants, look as though they were tossed down on the cave floor only yesterday.

In the deserts of coastal Peru and Chile some open-air sites are so dry that maize plants (along with almost all other organic materials) preserve just as well as they do in caves. In either case, the payoff is enormous. Archaeologists and paleoethnobotanists can collect large samples of uncharred plant material and study the characteristics of the plants in detail. This is preferable to working with charred plant remains because the charring process, even though it allows for the preservation of some plant parts, usually damages and distorts the material so that shape and size characteristics are significantly altered, and in some cases the more delicate parts have been entirely burned away. In contrast, uncharred plant remains can shrink as they lose their water content, and their original colors may fade, but in most other respects they remain unchanged.

MAIZE MACROREMAINS

Efforts to collect and analyze maize macroremains have paid off handsomely over the past several decades. The remains have helped botanists answer many questions about maize's evolution and spread: What did the earliest maize look like? What was the rate of change from the earliest varieties to the later forms? What varieties were present in different parts of the Americas? How did maize's productivity change over time? This set of questions can only be answered with large collections of maize, and, fortunately, some cave sites have yielded just that.

Paul Mangelsdorf was one of the first botanists of the modern era to analyze the large archaeological collections of maize from dry caves in the US Southwest and Mexico. He describes the history and results of his many years of maize studies in his book *Corn: Its Origin, Evolution and Improvement*. Even more detail is provided in the innumerable monographs and papers that summarize the archaeological studies of which his botanical analyses were a part.[5] Mangelsdorf's discoveries led to the standard view of maize development that we have today: that the earliest domesticated maize was relatively small and comparatively less productive than modern maize and that human selection over countless generations eventually transformed it into more than three hundred varieties, most of which have larger ears, in both length and diameter, with larger and more numerous kernels on more rows.[6]

As we saw in chapter 3, Mangelsdorf and R. G. Reeves hypothesized that the earliest domesticated maize descended from an extinct wild maize that had derived some of its characteristics from an introgression (or crossing) with the wild tropical grass *Tripsacum*.[7] Others, especially George Beadle, long contested this theory and instead championed the hypothesis that maize was domesticated teosinte.[8] As we will see in chapter 8, the maize-from-teosinte hypothesis has been confirmed by powerful new genetic analyses, and Beadle's estimate of the number of key genetic mutations that took place during the domestication process (about five) has proven to be largely correct. Morphological analyses of macrobotanical samples of maize have provided another important way of tracing its emergence and evolution. One thing that Mangelsdorf and Beadle agreed on was that once early domesticated maize appeared on the scene, it continued to increase in size and productivity as it became progressively more important, both practically and symbolically, among ancient Americans.

In recent years botanists have been able to document the specific similarities between maize and teosinte and measure the morphological characteristics that have changed through selection during the course of domestication.[9] In addition to undertaking these important reanalyses, archaeologists and paleoethnobotanists are recovering more samples of maize from ancient sites. As more collections become available for study, they expand the primary evidence for maize's uses and its dispersal. Let's look at some of the results of the new analyses of maize morphology and what they tell us about the plant's history during the past several thousand years.

TEOSINTE BECOMES MAIZE

To tell this story we need to look again at the earliest samples of maize macroremains discovered so far: three tiny, uncharred cob fragments found by Kent Flannery and his team during the 1966 excavations at Guilá Naquitz Cave in the Valley of Oaxaca.[10] The archaeobotanist Bruce Benz, who conducted the most recent study of the ancient Guilá Naquitz maize morphology, has found that the three cobs can tell us a great deal about the nature of early human-teosinte interaction.[11] The first characteristic of all three specimens is that they are indeed cobs (see figure 1.2—the Guilá Naquitz cobs). That is, the grain was attached to a **rachis** (comprised of many individual segments, called rachids) that did not break apart upon ripening. The rachis is the part of the inflorescence that attaches the grain segments to one another so that they stay together as a unit until ripe, forming a spikelet or ear that protects the seeds. In wild grasses, the layer of cells, called the **abscission layer,** that attaches each rachid segment to its neighbor, is relatively thin and genetically programmed to stay rigid only while the inflorescence is green and immature. This allows the seeds to fill out with sugars, which eventually convert into starch. The important thing about this rachis structure in the wild is that, when mature, the individual seeds (or grains) will disarticulate naturally as the abscission layer between adjacent seeds breaks down, allowing them to disperse and spread more widely as they fall to the ground. This natural dispersal mechanism gives the seeds a chance to germinate without crowding one another, thereby increasing the probability of successful reproduction. Benz found that the Guilá Naquitz cobs had already undergone selection for a non-disarticulating rachis; in other words, they had a rigid rachis with an abscission layer that didn't break down when the seeds were ripe, so the ear could not shatter after maturing. That is in fact what makes them "cobs." This is a very common characteristic of most domestic grain and other seed crops, including wheat, barley, and rice.[12]

Benz concludes, however, that the cobs are still quite teosinte-like in several respects. All three of the specimens are **distichous**—that is, they have two ranks of seeds, one on either side of the ear, alternating with

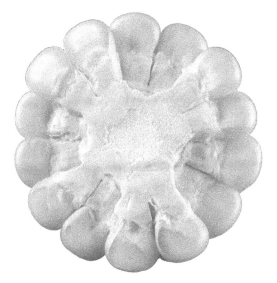

FIGURE 5.1. Transverse cross section of a typical ear of maize. This commercial variety has seven pairs of kernels in fourteen rows. The basal glumes join together toward the core of the cob, or the rachis, showing the symmetrical pairing of each set of two adjacent kernels, which are slightly offset from their neighboring pairs. The soft inner pith of the cob can also be seen—this is the vascular channel that supplies nutrients to the kernels as they grow and ripen to maturity before finally hardening and drying. (Photograph by Michael Blake)

one another (see the cob on the left in figure 1.2). Two of the specimens are two-rowed (one row per rank), and one specimen is four-rowed (two rows per rank). The grain-bearing spikelets are paired on the four-rowed specimen, meaning that it has two rows per rank like all modern maize. Even so, this specimen is still teosinte-like in that it is distichous. Modern maize is **polystichous** (which means that it has more than two ranks; modern maize normally has from four to ten ranks, but it can have up to twelve ranks) and has paired spikelets; this is why maize always has an even number of rows, ranging from eight to twenty, depending on the variety.[13]

The next time you find yourself holding a cob of corn, take a close look at the part of the kernel that is attached to the cob. You'll see that each kernel is part of a pair and that the base of each is tightly joined with its mate at the point where it attaches to the **cupule** (figure 5.1). In

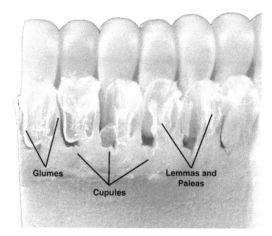

FIGURE 5.2. Longitudinal cross section of the same ear of maize shown in figure 5.1. The glumes embrace the kernels and are themselves nestled in the cupules, which form the hardest part of the rachis. Within the glumes and surrounding the base of each kernel are feathery lemmas and paleas—the soft parts of the cob that so frequently get caught in one's teeth when eating corn on the cob. (Photograph by Michael Blake)

teosinte and a variety of maize called pod corn, these cupules almost completely surround and protect the seed, whereas in maize they have retracted to form the hard, woody part of the cob—the part that stops your teeth when you bite down on the tender kernels of corn. Surrounding the base of each kernel where it is nestled in its cupule is the chaff—a set of thin, feathery bracts called lemmas and paleas, which are in turn surrounded by the harder bracts, or **glumes** (figure 5.2).[14]

The other teosinte-like characteristics of the Guilá Naquitz cobs are their texture and color. Two are mottled and smooth, have shiny surfaces, and are very hard in comparison to modern maize cobs. They are only like maize in that they have rigid rachises and grain-bearing spikelets that are perpendicular to the rachises. The third cob—the one with four rows—is more maize-like in that, in addition to these two traits, it has paired spikelets. All three cobs have glumes that are harder and much longer than the glumes in modern maize. In teosinte the glumes are much harder still and form a very durable fruitcase that completely protects the grain after it falls to the ground. By the time teosinte made its way to Guilá Naquitz, the people who domesticated it had already selected for the mutation that led to a naked grain that was only partially surrounded by a much softer glume—although it was still not as soft or as short as in modern maize.[15]

After completing his morphological analysis of these cobs and comparing them with maize cobs from the Tehuacán caves, Benz concluded that teosinte was already domesticated but that the genes that were under selection were not yet entirely fixed. One of the fascinating implications of these ancient teosinte-like domesticated maize plants is that people were selecting for characteristics that made the seeds easier to harvest and consume—as indicated by the rigid rachis forming a cob and by the naked grain, unprotected by the hard fruitcase found in teosinte. How they discovered the initial mutations that led to these developed traits is still a source of debate and conjecture, but there seems to be little doubt that these were the traits that people continued to encourage and foster during the subsequent millennia.[16]

INCREASING MAIZE PRODUCTIVITY

The path from the tiny, teosinte-like maize ears that were grown in the Valley of Oaxaca more than 6,000 years ago to, as George Beadle put it, the "man-made monstrosity" that is modern maize, goes through the Tehuacán Valley, some 180 kilometers north of Oaxaca.[17] Analyses of the macrobotanical remains of maize have helped document the rate and magnitude of change in traits such as increasing kernel size, kernel number, cob length and diameter and decreasing glume size. All of these traits were visible targets of selection that early maize farmers could decide for or against, and their evolution can be relatively easily plotted by taking careful measurements of the macroremains.

Unlike at Guilá Naquitz Cave, where only four cob fragments were found in the Archaic period deposits, the caves in the Tehuacán Valley yielded more than 24,000 well-preserved cobs and other maize-plant parts. Paul Mangelsdorf and Walton Galinat described the characteristics of these maize macroremains in great detail and were able to outline the major changes in maize ear morphology over the course of about 6,500 years.[18] Mangelsdorf was much concerned with classifying the Tehuacán Valley maize into types that demonstrated the evolutionary sequence from early wild maize to later domesticated varieties and tracked the incredible growth of the maize ear through the millennia. His main classes were Wild, Early Cultivated, Early Tripsacoid, Nal-Tel-Chapalote Complex, Late Tripsacoid, Slender Pop, Zapalote Chico, Conico, Dent, Tepecintle, and Chalqueño. He arranged these varieties by their changing percentages through time, which yielded a chart that showed when a variety appeared, how it increased in popularity, and how it either declined and disappeared or expanded to become dominant in the later periods. As we saw in the last chapter, Mangelsdorf and MacNeish also created a now-iconic image of maize's evolution by lining up a series of cobs, from the smallest and earliest on the left (which they classified as wild maize) to the largest and most recent on the right—visually demonstrating maize's march of progress to increasing productivity (see figure 4.2). This famous image gives the false impression that the relatively small "wild maize" begat the early cultivated variety of maize and was not itself the result of a long process of selective development. Teosinte is not in the family picture.

In spite of this omission, the Tehuacán maize family portrait is still widely reproduced because it represents the basic evolutionary history of the one of the most important crops domesticated in the New World. However, most botanists now agree that the portrait needs some minor modifications. One key change would be the addition of teosinte—in particular, *Zea mays* ssp. *parviglumis* and *mexicana*. Indeed, as we will see below, DNA analyses have now shown that teosinte is the original wild maize and that Mangelsdorf's "wild" maize is in fact early domesticated maize. Another change would be to add some spacing between the specimens to show that the evolution of the maize ear under domestication was not actually a continuous and constant growth, but was instead a process punctuated with variations in the rate of change and tempo of selection. Not all traits changed simultaneously.

Much of our sense that the Tehuacán maize family portrait needs some tweaking is based on morphological analyses carried out by Hugh Iltis, Bruce Benz, and their colleagues and students over the past thirty years. Benz, for example, has been systematically remeasuring samples from the Tehuacán maize collection and directly dating large numbers of specimens where possible to try to establish a more precise timeline for the observed morphological changes. He has discovered that there have been significant changes in the rate of evolution of several of maize's morphological traits. His analyses are still underway, but some of his published results show that maize's evolution is much more complicated than was previously suspected.[19] In one of his more recent analyses, Benz examines the maize remains from El Riego Cave in the Tehuacán Valley, showing that cupule width, grain width, row number, and cob size all have periods of change and stability—sometimes independent of one another (figure 5.3). By comparing this detailed sequence of change over the 4,000 years of the cave's occupation, Bruce Benz and his colleagues were able to make inferences about selection processes and variations in the rate of evolution of individual traits. Here's how they did it.

Benz and his colleagues selected El Riego Cave for reexamination because it had well-preserved maize remains that had not been systematically studied or dated by the MacNeish-Mangelsdorf team back in the 1960s and 1970s. The original excavators had meticulously excavated the cave deposits, carefully recorded the superimposed sequence of individual cultural layers as they went, and screened all the soil to recover artifacts and botanical remains. This meant that the maize samples they found could be placed in a stratigraphic sequence and then compared with one another, from the earliest to the latest of the nine layers in the deposits—which spanned about 4,000 years. But rather than rely on the stratigraphic sequence to determine the ages, Benz selected thirty maize cobs for AMS radiocarbon dating, twenty-four of which returned successful dates. The oldest specimen was between 3617 to 4864 cal BP, and the youngest was only 54 to 317 cal BP. Benz cautions that some of the directly dated older maize cobs had moved upward into more recent layers, while some of the younger cobs had moved downward. As we saw in the previous chapter, this mixing, probably due to both human and rodent activity in the cave, is significant enough that one must question the age determinations of the deposit's layers that were made prior to the use of AMS dating. Benz and his team overcame this problem by looking at the variation in maize cob measurements only for AMS-dated specimens: "the AMS dates provide us with an absolute chronological basis of morphological comparison largely independent of stratigraphic mixing."[20]

They examined row numbers, cupule width (a proxy for kernel width), rachis diameter (cob diameter), and rachid length (an estimate of kernel thickness). Rachid length remained relatively constant, between about 3.0 and 3.5 millimeters, during the entire period of occupation. The overall diameter of the cobs also remained fairly constant, between about 4200 and 3000 cal BP, and then increased steadily until about 1000 cal BP, after which the cobs narrowed almost to Formative period sizes. Benz concludes that when all the measurements are looked at together there is very little increase in productivity of the El Riego maize from around 4200 to 2290 cal BP. It isn't until about 1600 cal BP that maize cob productivity, in terms of grain size and quantity, increased appreciably. Following 1600 cal BP, people seem to have encouraged

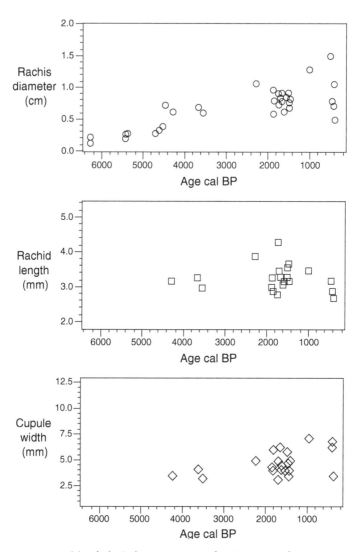

FIGURE 5.3. Morphological measurements of maize remains from individually dated maize cobs, spanning more than 4000 years of occupation at El Riego Cave in the Tehuacán Valley, Mexico. The measurements, part of ongoing work by Bruce Benz, show that there are gradual increases in cob (rachis) diameter (top graph) and cupule width (bottom graph), with a marked increase beginning about 4,500 years ago and then another increase about 2,000 to 1,500 years ago. In contrast, rachid length does not show such a strong change over time. (Illustration by Michael Blake, redrawn from Benz et al. 2006:78, Figure 5-2)

more variety in maize: there were some larger and other narrower cobs, some bearing wide kernels and others bearing much smaller kernels in more rows. This kind of "fine-grained" analysis, so to speak, while still dealing with relatively small sample sizes, is important because it shows that the lineage of maize is not simply a one-way, continuous progression to ever increasing ear sizes and productivity. In other words, after the Formative period, as people became more dependent on maize as a staple crop, they also became increasingly concerned with the plant's diversity. I think it quite likely that people were attempting to create and encourage maize varieties with different characteristics that would have the potential to grow in different conditions and produce crops with an ever wider range of traits—perhaps including resistance to droughts, pests, increased preservability, and varying grain qualities.

AGRICULTURAL SELECTION—BEYOND PRODUCTIVITY

We tend to think of agricultural selection in terms of increasing productivity—that is, increasing the yield of the plant part that is of prime economic interest, and that usually means food content. While this is undeniably important for many crops, it is not always what farmers are most interested in, and even if it is, they often concern themselves with other characteristics of the plant as well. In our modern agro-industrial society, consumers are often concerned with a food's appearance, texture, and flavor—even when these traits contradict one another. Many consumers in North America and Europe, for example, have a difficult time finding the flavorful plant foods that they remember from their youths. New varieties of fruits and vegetables that have been introduced in the past thirty years look appealing in the grocery store but are almost completely devoid of flavor. It seems that the more flavor-filled varieties are less visually attractive, have shorter shelf lives, and can't travel far without spoiling or bruising. Large wholesalers and retailers are loath to stock varieties that, while they might be more palatable, are less salable because they don't look as attractive. In recent years, however, there has been a noticeable backlash among some consumers. People are more willing to search out and buy varieties that have more flavor and other qualities that had all but disappeared in many urban settings. Now we can visit organic supermarkets and farmers' markets and even grow our own fruits and vegetables in urban gardens to obtain foods that have a range of culturally important qualities beyond the one or two commercial attributes that are thought to be important by agro-industrial producers.

The cultural preferences of ancient maize producers likely had a significant effect on the selection and rates of evolution of the different traits and characteristics that show up in the enormous range of maize varieties found throughout the world today. Varieties of maize are usually called landraces, or simply races, and they are well known by both the farmers who grow them and the botanists who study them. A remarkable series of books published as part of the massive Races of Maize study in the 1950s and 1960s, with titles including *Races of Maize in Mexico* and *Races of Maize in Peru* (along with nine other volumes for every region in Central and South America), describes in detail all the different races of maize in the Americas, providing the first comprehensive descriptions of all of the variation in maize that was known at the time.[21] These works were intended to document the relationships among the different varieties and correlate the morphological variation in maize with both environmental and cultural factors. It is clear from the enormous variation in many aspects of maize—the reproductive structures in particular—that people, in concert with the environments in which they lived, selected for a wide array of traits that went well beyond mere food productivity. They were interested in grain color, cob shape, kernel shape and size, starch texture, and so forth (figure 5.4). It is estimated that there are some hundred races of maize, with approximately thirty-two originating in Mexico and another forty-eight in Peru. The others come from all over the Americas, from Canada to Chile.

As we saw in some of Benz's studies, there were two clear processes that began at least 6,000 years ago: the size of the maize ear generally increased over time and the variety of shapes and sizes increased over time. What did this mean for the farmers who grew maize? Were they trying to create varieties that were more productive (which had been one of the goals of the botanists who initiated the Races of Maize project)? Were they interested in developing forms and flavors and characteristics that were aesthetically appealing, quite apart from nutritional value? Or were they doing some of both? I suspect the latter. I believe that the general size increase of the ears over the millennia was part of the attempt to increase productivity while the remarkable variation in ear shapes, kernel colors, kernel textures, and starch properties was more of an aesthetic quest.

In a recent article Bruce Benz, Hugo Perales, and Steven Brush report on an ethnobotanical study they undertook of variation in the maize of the Tzeltal and Tzotzil Maya farmers of Highland Chiapas. Their work shows that there are important aesthetic considerations that are of great

FIGURE 5.4. Ears of maize in the International Maize and Wheat Improvement Center's collection show the diverse colors and shapes found among modern maize varieties. (Photograph by Xochiquetzal Fonseca/CIMMYT)

concern to present-day Maya maize growers. For these farmers, however, visual traits such as grain color are not simply aesthetic. Rather, they can represent spiritual, sacred traditions linking people and their most important plants to the supernatural forces that created all life. Ethnobotanists working with these farmers in the 1970s found that the Tzeltal have five color varieties of maize representing several distinct races and hybrids, while the neighboring Tzotzil make as many as fifteen racial and hybrid distinctions.[22] When Benz, Perales, and Brush met with groups of farmers from the Tzeltal and Tzotzil communities and asked them to sort examples of maize into different varieties that they recognized, most were able to distinguish their own varieties of maize from those grown in neighboring communities. The most important factor for them in differentiating between the examples they were presented with was color. The researchers found, however, that there were no distinct physiological or adaptive traits that correlated with the groupings the farmers made—instead the varieties that the subjects identified with the most consistency were mainly those that were products of their local practices of planting seeds passed down through the generations, which led to recognizable variants.

A similar process of maize variation has been observed among the Yucatec Maya farmers of Yaxcabá, east of Merida in the northern Maya lowlands. John Tuxill and his colleagues have been studying the farming practices of this community over the course of many years and have found that comparable principles are applied to maize farming and classification systems. There, farmers recognize at least sixteen indigenous maize varieties, and the attributes that are most commonly used to distinguish these varieties from one another is the length of time it takes the plants to mature and the shapes of the ears.[23] The fastest maturing variety—called *nal t'eel,* meaning "rooster maize"—takes only seven weeks to progress from seeding to harvest, and the longest maturing variety—called *x-nuuk nal,* or "old maize"—can take up to four months. Within these varieties, however, the most common characteristic used to distinguish individual populations is the color of the kernels, which can be yellow, white, dark purple, or blue-tinted. Tuxill and his colleagues discuss how the farmers use the color characteristics to determine which seeds will be kept for replanting the next crop and which varieties have the most promise for desirable traits linked to productivity, reliability, and resistance to pests. But the sacred, symbolic dimension of maize's color underlies some of the subtler aspects of planting choices, along with the ultimate use of the kernels for consumption. As Tuxill and his team put it: "The prevailing kernel colors of Mayan maize varieties—red, yellow, purple-black, and white—gave maize an additional level of symbolic meaning and significance in religious offerings and other ceremonial occasions."[24]

While the visibly recognizable varieties of maize grown throughout the Americas can be linked to environmental adaptation, it is clear that the sacred and ceremonial dimensions attributed to the plant also played, and continue to play, a role in people's selections of ear shape, kernel color, flavor, and consistency, along with many other properties. The plant's genetic flexibility and adaptability made it possible for people to grow it in many different regions far removed from its original homeland and allowed them to continuously select for their preferred variations. Thanks to the fortuitous cases where ancient examples of maize macroremains displaying these preferred characteristics are well preserved, it is possible for us to see the result of this selection process as it played out over countless generations—both of farmers and of the amazing plants that sustained them.

CHAPTER 6

Maize through a Microscope

Microremains

On January 16, 1831, three weeks after leaving Devonport, England, the HMS *Beagle* dropped anchor off Porto Praia on Santiago Island, one of the Cape Verde Islands off the west coast of Africa. The insatiably curious young Charles Darwin, still in his twenty-first year, wrote in his journal that the air was often hazy with a fine brown-colored dust. The morning before anchoring his curiosity got the better of him, so, naturally, he collected some of that dust by shaking it out of the gauze of the ship's weather vane. He was not the only dust collector in those waters: the great British geologist Charles Lyell had also collected several packets of North Atlantic dust during his voyages. What is most remarkable about Darwin's dust, however, is what he did with it after returning to England five years later. He sent the packages to an illustrious German biologist named Christian G. Ehrenberg, one of the pioneers of microscopy (and the person who coined the term "bacterium").[1] Darwin noted that, within the dust, Professor Ehrenberg identified particles of "infusoria with siliceous shields, and of the siliceous tissue of plants. In five little packets, which I sent him, he has ascertained no less than sixty-seven different organic forms! The infusoria, with the exception of two marine species, are all inhabitants of fresh-water. . . . although Professor Ehrenberg knows many species of infusoria peculiar to Africa, he finds none of these in the dust which I sent him: on the other hand, he finds in it two species which hitherto he knows as living only in South America."[2] These siliceous infusoria were of course what we now call phytoliths—and it would be wishful thinking to hope that at least some of them might have been maize infusoria blowing on the winds over the Atlantic. It is unlikely, though, since the prevailing trade winds at the latitude of the Cape Verdes blow from the northeast to the southwest, not the reverse. Ah well, the discovery of maize phytoliths would come later.

Phytoliths are one of three types of very small-scale plant parts, along with pollen and starch grains, that have become increasingly important in our understanding of ancient environments and human-plant interactions. Because they can only be observed and studied at the microscopic level, they require a special expertise—and patience—to understand. Even for those of us who are not botanists or agriculturalists, it is relatively easy for us to identify a maize plant when we see it. Maize has a distinct form, or "habit," a term botanists use to refer to the overall shape of a plant. In most places where maize is grown today—that is, far removed from the tropics—there are few other grasses that can claim to be "as high as an elephant's eye."[3] And, of course, even more striking is the ear of maize— no other type of grass produces such an appendage. But few of us know that maize has unique features on the micro-level as well, features that can only be observed with the aid of a high-powered microscope.

As we glimpsed in chapter 4, the ability to identify maize microremains has transformed the view we have of maize's evolution and spread. Let's now look a little more closely at the three distinct types of maize microremains that have been identified in the archaeological record: pollen, phytoliths, and starch. These plant parts have properties that are quite different from the plant's larger structures. Macro plant parts, including stalks, leaves, seeds, flowers, and roots, are not particularly durable when left out in the open or when buried in the ground. Most of this sort of vegetative material decays rapidly, depending on climate and soil conditions. In contrast, pollen, phytoliths, and even starch granules can be preserved for much longer spans of time, even in conditions that lead to the decay of the larger plant parts that produce them. Researchers have known about this for a long time, and for many decades pollen has been successfully identified in archaeological and environmental sites as a way of inferring the presence of various plant species. The archaeological applications of phytolith and starch granule identification are much more recent—yet they are proving to be just as revealing of past plant use as pollen. Before discussing some of the recent discoveries that microremain analyses have yielded, I will introduce the main characteristics of each type of material. This requires a

FIGURE 6.1. Many plants, including maize, take advantage of the wind to disperse their pollen and ensure reproduction. Various species of coniferous trees, such as pine, are famous for producing vast amounts of pollen that can travel enormous distances. This photo shows a pollen cloud being released into the air when a person brushes the branch of a pine tree on a windy day. (Photograph by Tangopaso, Wikimedia Commons)

brief foray into the world of botany—well worth it for readers like me, who are not experts in the field.

POLLEN

Pollen is essentially the plant world's equivalent of sperm cells. Pollen grains are produced in a plant's anther and are composed of reproductive cells, including the male gamete, that are surrounded by a durable outer casing, or **exine,** made of **sporopollenin.** The exine often has a distinctive surface shape that can be identified to family, genus, and even species. Many plants produce pollen in huge quantities.[4] The types of plants that produce the most pollen usually rely on the wind to transport it over vast distances (figure 6.1). Others that produce less pollen rely on insects to pick up the pollen grains and transfer them from plant to plant, or, in the case of self-pollination, from the male to the female part of the plant. Members of the grass family (Poaceae) rely on windborne pollination and therefore often produce prodigious quantities of pollen. This is one reason that grasses don't have large, attractive flowers—they don't rely on insects to move their pollen. Pollen grains are tiny, ranging in size from 15 to 100 microns (μm—a thousandth of a millimeter). Species that rely on the wind for pollination produce smaller, more aerodynamic pollen shapes, while those that rely on insects produce larger and

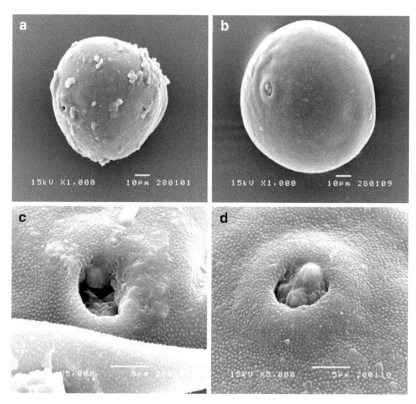

FIGURE 6.2. High-magnification images of maize pollen produced by a scanning electron microscope. (Photograph from Chantarudee et al. 2012)

more irregularly shaped grains that are often sticky so that they adhere to the insects that transport them. As members of the grass family, teosinte and maize are odd in this regard. Even though they rely on windborne pollination, they have much larger pollen grains than most other grasses (figure 6.2).

The distinctive shapes of individual pollen grains form as part of the exine layer, or outer sheath of the grains. Unfortunately, the pollen grains of most grasses are indistinguishable from one another—their shapes and surface traits are very similar regardless of species because all grasses evolved relatively recently and have not had time to develop individual shapes and surface characteristics. Even so, grasses are ancient, and the fossilized pollen grains of a primitive grass, *Monoporites annualtus,* have been identified in deposits dating to the early Palaeocene period, which began roughly sixty-five million years ago.[5]

The one characteristic that distinguishes maize pollen from most other grasses is its size. Maize produces pollen grains that are among the largest of any New World grass and so can be easily spotted in pollen samples. In fact, maize pollen can reach upward of 125 microns in diameter, more than twice that of most wild grasses. To be a bit more specific, the pollen of modern maize hybrids ranges in diameter from 94 to 119 microns (mean 106±4.7 microns), while teosinte pollen ranges from 65 to 95 microns (mean 85±3 microns).[6] Maize pollen's larger size means that it can't be borne as far by the wind as pollen from teosinte or other wild grasses. To put it in perspective, one team of crop scientists experimenting with the properties of maize pollen shedding found that although maize pollen could theoretically travel up to 32 kilometers from its source (providing wind conditions were just right), very little of the pollen they tracked traveled more than 100 meters, and the bulk of it landed within 1 meter of its source.[7] In contrast, pine trees are famous among pollen experts (who are known as palynologists) for producing vast quantities of pollen that can travel for hundreds of kilometers. As we will see, these characteristics of maize pollen grains play a significant role in the story of maize's spread and its importance throughout the Americas.

PHYTOLITHS

Unlike pollen, phytoliths are not reproductive organs. They are microscopic silica particles produced by the deposition of silica inside plant cells (almost all parts of a plant can produce phytoliths) and often in the spaces between cells as well. Not all plants produce phytoliths—a point I will come to later. Silica is taken in as naturally occurring monosilicic acid through the root systems of some plants and accumulates in the plant during normal growth.[8] Although not all of their functions are well known, phytoliths may be analogous to bones in vertebrates—they can provide the strength and rigidity that helps plant structures, such as leaves, stalks, and fruits, maintain their forms.[9] Some researchers have also proposed that the silica particles in the epidermal cells help to toughen up the surface of plant structures to defend against pests, such as fungi and herbivores. Phytolith formation is controlled partly genetically and partly through a plant's adaptation to local soil and climatic conditions.

Phytoliths generally overlap in size with pollen grains, ranging from 5 to 100 microns, although in grasses they average around 10 to 20 microns. Usually only one phytolith is produced within a single cell, but they occur in the millions in any given plant—with the exact number depending on the size of the plant and the total number of cells. These microscopic particles are extremely durable—much more so than pollen under most circumstances. Since phytoliths are almost pure silica (with a few other elements mixed in, including carbon) they are inorganic and do not decay. Therefore they can last for millions, even tens of millions, of years.[10]

As with pollen, the shapes of phytoliths vary widely across the plant kingdom, and particular shapes have been associated with individual species and even with specific parts of the plant (e.g., with stalks, leaves, and fruits). Phytoliths are generally translucent and three-dimensional, falling into general shape groups that specialists are able to identify under high-powered magnification (figure 6.3). These characteristics have long been known—especially in Europe, where much of the earliest work on phytoliths was carried out, beginning in the 1800s. In the New World, studies of phytoliths in archaeological deposits has been more recent, but it has been particularly important for understanding the origins of New World agriculture—maize agriculture in particular. Maize and teosinte phytoliths have similar shapes to those produced by other members of the grass family, but even so, they have distinct characteristics that can help to distinguish them.[11]

STARCH

Like pollen and phytoliths, starch grains (or granules) are produced by some plants in great abundance, and their shapes and sizes may vary from species to species. Almost a century ago a biologist named Edward T. Reichert published a book describing the differences in the characteristics of starch produced by various species of food plants.[12] Of course, we humans have long been interested in starch because our lives depend on consuming it daily. But we learned to identify it at archaeological sites only relatively recently. Starch granules are the "fuel" that many plants need to reproduce and survive. They are produced as two types of sugar—glucose polymers (called amylose) and glucose polysaccharides (called amylopectin). Each is composed of chains of strongly bonded oxygen, hydrogen, and carbon atoms.[13] Starch granules are highly efficient mechanisms for storing excess glucose inside cells so that it can be broken down for later use. For example, during germination,

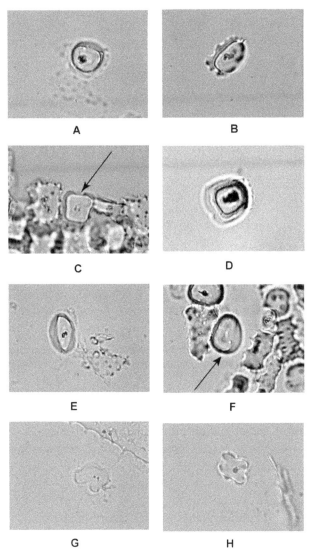

FIGURE 6.3. High-magnification images of *Zea* phytoliths, including maize and teosinte. (Photograph by Robert G. Thompson, in Hart et al. 2011:7, figure 6)

FIGURE 6.4. Sonia Zarrillo scraping charred organic residues from a ceramic fragment recovered at the Santa Ana-La Florida site in Ecuador. The residues yielded maize starch grains and a radiocarbon sample for dating. (Photograph courtesy of Sonia Zarrillo)

a seedling will draw on the energy it has stored as starch to ensure cell growth until it forms mature plant structures—such as leaves, roots, and stems—that can extract energy from soil and sunlight. When humans harvest and process the parts of plants that contain abundant starch—such as roots, tubers, seeds, and some stalks—starch granules are released, some of which (those that are not eaten) might eventually become part of the future archaeological record (figure 6.4).

As with pollen and phytoliths, starch granules are very small (ranging from 3 to 100 microns in diameter). But unlike pollen and phytoliths, they are not nearly as durable in the archaeological or environmental record. The main reason for this is that they can be food for microorganisms—and in the never-ending quest for life microorganisms are not likely to let good food go to waste. Even so, under some circumstances—such as the extreme aridity of a cave site or the physical protection provided by a microscopic crevice in an ancient tool—some starch granules survive in the archaeological record, as we will see below.[14]

ANCIENT MICROREMAINS

Before we take a closer look at some specific archaeological examples of pollen, phytolith, and starch evidence for how the domestication and spread of ancient maize took place, we should address some of the questions and criticisms raised over the years that are driving interesting new research that is helping to make the interpretations of the history of ancient maize much clearer. While no one doubts the utility and importance of microremains studies, some heated debates have arisen on the topic during the past few years. These debates provide an important illustration of how archaeological science is carried out and show the link between expectations, data collection, and interpretations. One key issue concerns the difficulty of dating microremains, which, until recently, usually had to be dated by association. As I mentioned in our brief introduction to the dating of macroremains in chapter 4, it has at times been difficult to assume that the dated charcoal or other organic materials associated with a set of microremains are really of the same age, since either the dated material or the microremains could have moved through stratigraphic layers due to human-induced or naturally occurring disturbances.

Archaeologist Kent Flannery has cautioned that it can be difficult to be certain that artifacts and other remains found at archaeological sites are what they seem—or rather, their associations may not be what they seem: "A packrat may bury a younger cob of maize in a rodent hole faster than it takes an archaeologist to turn around."[15] Similar processes could also operate on microremains. Earlier I mentioned the problems of root action, rodent and insect burrowing, and other processes of earth movement—even some carried out by humans. This is why it is necessary to use multiple lines of evidence in trying to interpret the archaeological remains from any site. If we find all three types of maize microremains in multiple archaeological contexts at a site, and they are all more or less independent of one another, then it is more likely that maize was present at the site during the time period represented by the layer or deposit in question.

For example, at the site of Real Alto in Ecuador, which I discussed in chapter 4, the recovery of both phytoliths and starch granules lodged in the crevices of several milling stones on the floor of Structure 20, which was dated to the Valdivia III period, is compelling evidence of the presence of maize at the site. But how can we be sure that the milling stones are in fact associated with the house? Milling stones are unlikely to move around much in archaeological deposits, especially in the absence of other signs of disturbance. So in the case of Structure 20 we have to assume that the milling stones are as old as the ceramics with which they are associated. Because of their particular styles of decoration and so forth, we can also assume the ceramics date to the period that the archaeologists say they do. Ideally we would like to have a radiocarbon date on charcoal scraped off one of the milling stones or potsherds. This would help to provide an independent line of evidence about the antiquity of the microremains. But in the case of Real Alto, this was not possible. Instead, the archaeologists looked for other ingenious ways to test the age of the specimens. They found that the types of maize phytolith shapes found at the site were characteristic of primitive forms of maize instead of being identical to forms found in modern varieties. While not proving that these phytoliths are ancient, this line of evidence is consistent with the expectations that the phytoliths, milling stones, and house structure remains are all closely associated with one another and likely fall within the age range of the Valdivia III period.

Paleoethnobotanists working in the humid tropics often find that microremains are the only part of a plant remaining after it has spent centuries and millennia buried in soil. Deborah Pearsall found that, at the Finca Cueva site, located in the Jama Valley of coastal Ecuador, charcoal fragments tended to become scarcer with the increasing depth of the deposits. She concluded that the longer charcoal fragments remained in the ground at open-air sites in humid lowland tropical regions, the more likely they were to decay and eventually disappear. This means that standards of plant identification and quantification that work in temperate or arid tropical environments may not always be useful in the humid tropical lowlands.[16]

Some researchers insist on finding the larger remains of plants to demonstrate domestication, because it is in these larger plant parts that one can clearly see the evolution of traits that were selected under domestication. However, in the case of maize one simply needs to find any trace of the plant—macro or micro—outside of teosinte's natural habitat in order to document domestication. So whether we are looking for maize in North or South America, microremains should be as good as macroremains to document its presence. But for this to work we must be certain the microremains are in fact from maize.[17]

Maize pollen is one of the most difficult to be certain of because, as mentioned earlier, it is so similar to the pollen of most other grasses. Maize pollen grains range in size within any given plant and among the different varieties. On average, maize pollen is larger than teosinte pollen, but their sizes do overlap. Unfortunately we don't have a clear picture of the rate of size increase in maize pollen during the plant's domestication history. We still have many questions to answer. How different is the size range between teosinte pollen and other wild grass pollen? How different is the size range of early maize from that of teosinte? Can we expect that maize pollen size increased at a constant rate through time, and if so, why? What is the evolutionary significance of increased pollen grain size in maize? These questions are important because much of the evidence for maize's early spread through the tropics comes from pollen sequences in which very small quantities of maize pollen—in some cases a single grain or two—are identified in sediment cores taken from lakes. And in these core sequences an interesting conundrum arises. The earlier the maize pollen is in the sequence, the smaller it is likely to be. But the smaller it is, the less likely it is to be exclusively identified as *Zea*—since size is the only reliable identifying characteristic.

Teosinte (*parviglumis*) pollen grains are on average 72 microns in diameter (although they can range from 63 to 80 microns)—roughly 30 percent smaller than modern maize hybrids—and they dry out 30 to 50 percent faster than modern maize.[18] The size of pollen grains influences two factors: the distance that the airborne pollen travels and the rate at which it desiccates. The larger and heavier the pollen grain, the shorter the distance it will travel in air currents but the longer it will resist desiccation. It may be that under domestication, maize planted closely together in fields could self- and cross-pollinate more effectively, so over time maize sacrificed the small size of its pollen, and thereby its mobility, for an increase in resistance to desiccation. Larger maize pollen grains had greater volume, which allowed them to hold more water, leading to more effective germination of individual pollen grains.[19] This might be especially advantageous because maize was planted outside of teosinte's home range and therefore experienced a greater variety of environmental conditions. The timing of the increase in pollen grain size should be genetically controlled and was likely a gradual evolutionary process like the other morphological changes maize underwent during domestication.

The range of grass pollen sizes is highly variable (20–55 microns) but a few taxa produce pollen that falls in the lower range of teosinte and so sometimes they cannot be distinguished from one other.[20] However, the size range variation of grass pollens is extremely helpful for distinguishing *Zea* pollen from other grasses and allowing researchers to distinguish between teosinte and maize. This has worked well in documenting the spread of maize. There are dozens of cases in which maize pollen has been identified in stratigraphic core sequences from lakes and swamps. Environmental scientists can often correlate the presence of maize with other evidence of agriculture—for example, increased charcoal fragments (indicating forest clearing and burning to make space for planting crops) and the increased frequency of maize phytoliths—in the levels where maize first appears. Sometimes maize pollen also correlates with the appearance of increased sediments in the column—which is a result of erosion caused by deforestation—and the appearance of pollen from weed species that occur in and around agricultural fields. These multiple indicators of agriculture help distinguish between natural causes of vegetation change (for example, changes in climate and natural forest fires) and cultural causes (such as agricultural forest clearing). Let's go to Guatemala to look at an example where the discovery of pollen and phytoliths in sediment samples from swampy habitats has helped to establish and date the early presence of maize agriculture in a region of the tropics.

Along the Pacific coast of Guatemala a series of mangrove swamps, canals, and estuaries have formed near where small rivers originating in the Sierra Madre Mountains descend onto the coastal plain. A team of archaeologists led by Hector Neff, Deborah Pearsall, Barbara Arroyo, and their colleagues sampled more than ten locations spanning almost 200 kilometers of coastline. They pushed metal tubes into the swamp sediments, pulling up core samples ranging from 1 meter up to 5 meters in length. They took the cores back to their laboratories, where they carefully opened them up so that they could see layer upon layer of estuary-bottom mud that had accumulated since the mangrove swamps first started to form. The researchers carbon dated organic material trapped in some of the layers, selecting well-preserved fragments of charcoal that had settled into the sediments, trapped by layers of accumulating mud. By selecting samples at different depths and dating them, it was possible to estimate a timeline, from the earliest layers at the bottom to the most recent layers at the top. In the case of one of their sample sites, Sipacate, they were able to date the earliest layers containing maize remains to about 5750 cal BP, and the layers continued, documenting another three thousand years. The researchers took very small samples of sediments (about one cubic centimeter) every centimeter along the length of their cores and scanned them for pollen, phytoliths, and charcoal—not just for maize, but for all identifiable plant species.[21]

The methods of sample preparation for both pollen and phytolith analyses are quite similar. First, the core has to be split lengthwise so that material for each type of analysis, including dating, is taken from exactly the same depth within the core. Then the splits are divided along their length into equal segments. Analysts usually divide the cores from top to bottom into 5-centimeter or 10-centimeter segments and take a single sample from each of these intervals. Neff's team sampled each centimeter along the core, giving them a very fine-grained look at the changes over the long time spans represented. Regardless of the sample interval—long or short—analysts have to be very careful not to contaminate any section of the sample with sediments from higher or lower in the sequence, since that could radically skew the results. Depending on the soil characteristics—whether it is very acidic or basic, whether it is composed of fine silts or larger particles, whether it has a good deal of organic matter or not—the sample is processed with acids, bases, or distilled water, and then fine screening/sieving, centrifuging, decanting, and drying. This rigorous set of treatments is aimed at removing as much of the soil sediments and organic material as possible so that the tiny pollen grains and microscopic phytoliths can be mounted on slides for viewing with a microscope.[22] In some cases analysts may use a scanning electron microscope (SEM) to get much higher magnifications of these minute botanical treasures.

Pollen and phytolith analysts will then scan a slide from each segment of the column, identifying and counting all the grains within the field of view until they reach a total of 200 or 250. After this total is reached, the relative abundance of the different species represented by the sample doesn't change much, so there is not much point going farther—except when researchers want to scan the whole slide to look for unique examples of rare species. But generally pollen and phytolith analysts are satisfied with 200 to 250 specimens per slide. After each type of pollen or phytolith is identified to family, genus, or even species, where such fine-scale determinations are possible, the counts are tallied, and their relative percentages are calculated. These are then plotted on a special graph devised to show the relative amounts of every category of plant from the top to bottom of the stratigraphic sequence. Most researchers also include a column for charcoal counts within the slide scan to help determine whether evidence exists for burning in the environment near where the sediments accumulated and, if so, how much.

Neff and his fellow researchers followed this general procedure for all the core sequences they analyzed in order to observe the changing percentages of pollen and phytoliths through time—grouping their counts into 5-centimeter or 10-centimeter levels, depending on how deep the cores were. Figure 6.5 shows the Sipacate 014 column pollen proportions and counts from a depth of 170 to 410 centimeters below the surface—spanning a period of about 5,500 years.[23] In the case of both the pollen and the phytolith sequences, the presence of maize is indicated very early on, by about 5300 cal BP, though in very small quantities compared with other, much more common wild species. Since the researchers also counted the particles of charcoal found along with the pollen, it is possible to see in most core sequences that the abrupt increase in the charcoal that was being washed into the swamp begins around the same time that maize pollen becomes more common. The appearance of charcoal suggests the increased frequency of forest burning, which, if not the result of natural fires, must have been the result of human manipulation of the landscape. The pattern at Sipacate 014 and the other sites that Neff and his teams studied is very similar to the pattern commonly observed in pollen cores from lakes and swamps throughout Central America and northern South America: the appearance in the cores of evidence of charcoal from forest burning corresponds with an increase in pollen and/or phytoliths from weedy species of plants, grasses, and in many cases *Zea mays* and other cultivated plants.

Lake core pollen evidence for the northward spread of maize does not appear so early. In fact, there are relatively few examples of lake cores as we move into the arid north of Mexico and the US Southwest—mainly because there are so few lakes. However, some examples have been found farther north in the United States and Canada. One of the earliest in the eastern United States comes from Lake Shelby, located along the Gulf coast of Alabama. There Miriam Fearn and Kam-Biu Liu found a single large grass pollen grain, identified as maize, in sediments from a core in the middle of the lake.[24] At a depth of 3.5 meters (that is, 3.5 meters deep in the core column extracted from the lake-bottom sediments) they found increased charcoal, suggesting burning, and were able to date a sample to ca. 3500 cal BP. This provoked a response from ethnobotanist Mary Eubanks, who cautioned that one grain of pollen was slim evidence indeed—and that it might have come from a different, but related, genus of grass.[25] Most would agree that one pollen grain from one core in one lake is not enough evidence to make a strong case for the presence of maize cultivation so early in the region.

One example from the far northern part of maize's range comes from Crawford Lake, a very small lake located midway between Toronto and

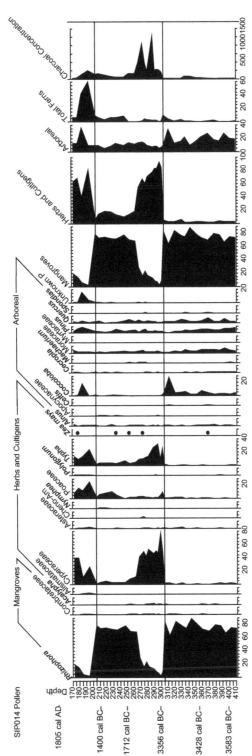

FIGURE 6.5. Standard pollen profile chart showing the changing frequencies over time of all of the identified species of plants represented in a sediment core sample. This sediment core, SIP014, from the Sipacate 14 location in coastal Guatemala, shows the appearance of *Zea* pollen grains at various depths below the surface (in the center of the graph, indicated by dots). (Illustration courtesy of Hector Neff, in Neff et al. 2006a:299, figure 5)

Hamilton, Ontario, just inland from the shores of Lake Ontario. Core sediment samples recovered as part of an environmental study of the fossil plankton abundance and geochemical characteristics of the lake during the past thousand years also uncovered pre-European-contact *Zea* pollen at a depth of up to 60 centimeters below the lake bottom. A series of samples from the core were AMS radiocarbon dated, placing the age of the maize at about 600 to 700 BP.[26] Many other examples of maize pollen and phytoliths from sediment gathered in lake and swamp locations away from archaeological sites, both in North and South America, can be found in the online Ancient Maize Map.

In addition to the environmental sites that have produced remarkable evidence of maize microremains, key sources of such data are to be found at archaeological sites—in the soils and on the artifacts found in and around the permanent and temporary dwellings of ancient Americans. We've already looked at a few examples, including the earliest, from Xihuatoxtla Rockshelter in Guerrero, Mexico (chapter 1), and the Archaic period cave sites in Panama (chapter 4). But let's take another look at the Real Alto site in Ecuador. One of the first people to recover and identify maize phytoliths from archaeological deposits in Latin America was Deborah Pearsall, who published a paper in 1978 describing phytoliths from house deposits at the Real Alto site (from the early Valdivia period, dating to about 4500–4000 cal BP) and the later site OGCh-20 (from the Machalilla period, dating to about 3500 cal BP) in coastal Ecuador.[27] When comparing modern maize with a series of modern local wild grasses, she discovered that maize had a higher proportion of cross-shaped phytoliths and that its cross-shaped phytoliths were significantly larger. She compared these results to archaeological samples and found that, out of the 1,700 ancient phytoliths that she examined, 32 were large cross-shaped types that could only have come from maize since they did not occur in any of the wild grasses she had studied. Pearsall's discovery was remarkable because it showed that, in the absence of macro-botanical remains of maize, it was still possible to find traces of maize plants at open-air sites—sites that could be securely dated using standard radiocarbon methods. This research has expanded greatly over the subsequent thirty years, and we now have a much more complete picture of ancient agricultural systems in general—and maize agriculture in particular—thanks to advances in phytolith research. During the past thirty-five years a great deal of new work on phytolith analysis has taken place, and the technique has been applied throughout the Americas. In combination with analysis of the other types of microremains, and other classes of archaeological data, it has proven invaluable, especially in cases where macroremains are not often preserved.[28]

In contrast to phytoliths and pollen evidence, starch grains are almost always only found in archaeological sites, where they can be preserved on tools, or occasionally in other contexts, such as on teeth. This is because, unlike pollen and phytoliths, starch grains have no hard mineral parts—they are composed completely of food, in the form of starch molecules, a substance that is equally as inviting to microbes as it is to humans. In one surprising example of the long-term survival of starch against all odds, archaeologists working on an archaeological site in the Solomon Islands reported finding plant starch granules trapped in the crevices of 28,000-year-old stone tools.[29] In the New World similar exciting discoveries are being made—though the materials are not quite as old.

Returning to the aforementioned Real Alto site on the coast of Ecuador, Deborah Pearsall and her colleagues recently studied a sample of milling stones from the floor of Structure 20. Experimenting with several recovery techniques, they extracted small samples of maize phytoliths and starch granules deeply embedded in the small cracks and fissures of several tools.[30] This discovery is significant because it provides an independent source of information about maize use at the site—a distinctly different context from the soil samples she had studied thirty years earlier. Remarkably Pearsall and her colleagues also note that the maize starch they found at Real Alto seems to be from two distinctly different types of maize—one of the granule types is found in hard-kernelled popcorn, and the other granule shape is more common in soft, "floury" types of maize.

In map 6.1 I've gathered together the earliest examples of dated microremain samples from each region, including both sediment cores from environmental sites—such as lakes, swamps, and lagoons—and samples recovered from artifacts and soils in archaeological sites. As in map 4.1 (which shows the distribution of directly dated maize macroremains and microremains) in chapter 4, these sites are connected within time range distribution zones—much as you might see on temperature range distributions for weather maps—based on their radiocarbon dates. In map 6.1 the ages of the maize microremains are mostly based on associated dates taken from dated organic material found in close proximity to the microremain, whether it be pollen, phytolith, starch, or a combination of the three. Many more sites and samples exist, but the ones included in these maps represent the earliest dates that have been recorded within a region (arbitrarily set to a 100 kilometer radius);

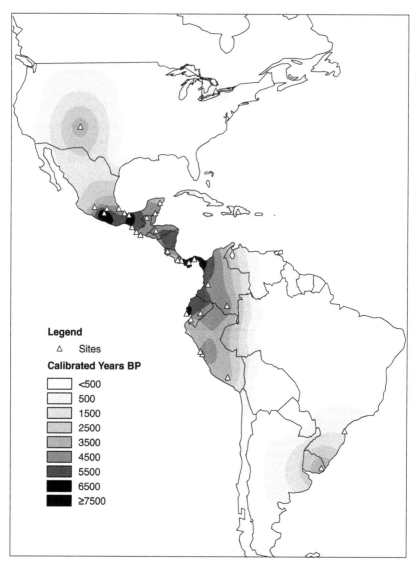

MAP 6.1. The age distribution of archaeological sites with indirectly dated maize microremains. The 1000-year-interval age distributions are based on the oldest date at a site. All sites located within a 50-kilometer radius of the oldest dated site in a region are excluded. (Created by Michael Blake and Nick Waber using sources referenced in Ancient Maize Map, http://en.ancientmaize.com/)

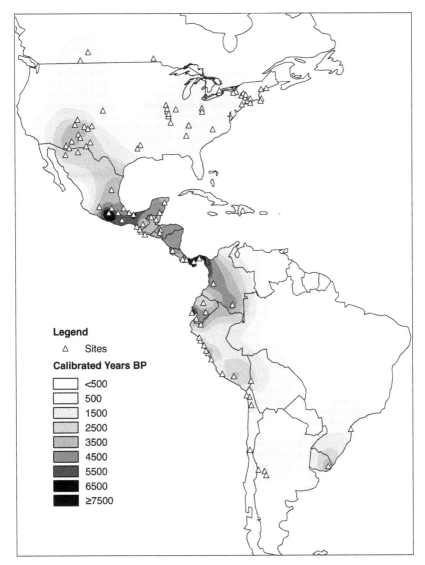

MAP 6.2. The age distribution of archaeological sites with both directly and indirectly dated maize macro- and microremains. The 1000-year-interval age distributions are based on the oldest date at a site. All sites located within a 50-kilometer radius of the oldest dated site in a region are excluded. (Created by Michael Blake and Nick Waber using sources referenced in Ancient Maize Map, http://en.ancientmaize.com/)

all later dates have been excluded. This map shows how microremains appear on the scene much earlier than macroremains and, when compared with the macroremains' age distributions, shown in map 4.1, illustrates the distinctive concentration of sites with ages over 6,500 years old spanning the region from Guerrero, Mexico (the homeland of maize's teosinte ancestors), across Central America to northwestern South America, as far south as coastal Ecuador. In contrast, map 4.1 shows that the earliest directly dated samples start in Oaxaca and more gradually spread northward and southward—leaving, however, a fairly large gap with no early directly dated maize in Nicaragua, Costa Rica, Panama, and Colombia. In Ecuador, the directly dated maize microremains are a few millennia more recent than the indirectly dated samples.

What does it look like when we combine the two maps? In map 6.2, we see a much clearer pattern, with a more continuous distribution of ancient maize across the zone from Guerrero, Mexico, in the north to Ecuador in the south. The northward spread of maize, from northern Mexico into the United States and eventually to southern Canada, mirrors map 4.1 quite closely because there have not been many older microremain samples reported yet. The southward spread of maize, at least to the equator, looks to be several millennia earlier than its northward movement beyond northern Mexico. The timing of the southern spread, south of southern Peru and Bolivia, is roughly the same as its northern spread. This suggests that it took maize longer to adapt to light and climate conditions north and south of the tropics than to new environments within the tropics—not surprising since teosinte is a tropical grass. It also suggests that the preservation of maize macroremains—and most other plant macroremains—in the hot, humid tropics is much less likely unless they can be found in perpetually dry caves.

In the next chapters we will look at some of the other archaeological indicators of maize's spread and at people's adaptations to this remarkably adaptive and peripatetic plant. We will explore two questions: How were people using maize and how was this miraculous plant evolving as it moved across continents?

CHAPTER 7

Elemental Maize

Tracing Maize Isotopically

Carbon is one the most basic building blocks of our bodies, and we get all of our carbon from plants. We obtain this crucial component of life either directly, by eating plants, or indirectly, by eating animals that have eaten plants. Plants, in turn, absorb carbon molecules from atmospheric carbon dioxide, taking in all of the three different "varieties," or **isotopes,** of carbon that exist in nature: carbon-12 (^{12}C), carbon-13 (^{13}C), and carbon-14 (^{14}C). The amounts of these different isotopes of carbon present in our own cells—in our skin, hair, bones, teeth, muscles, and so on—bear a direct relationship to the carbon we've consumed via all of the various foods that make up our diet. In other words, our bodies carry a carbon isotope "signature" that reflects, fairly precisely, what we've been eating during the course of our lives.

Carbon exists naturally as two stable isotopes and one unstable isotope. By far the most prevalent of the carbon isotopes is ^{12}C (it comprises 98.93 percent of all carbon on the earth), while the next most common is ^{13}C (comprising only 1.07 percent of all carbon). Both ^{12}C and ^{13}C are stable.[1] The third isotope of carbon, ^{14}C, is extremely rare in nature and gives off beta radiation, converting to nitrogen-14 (^{14}N) in the process—a topic we looked at in chapter 4. The values 12, 13 and 14 refer to the number of protons and neutrons surrounding the nucleus of a carbon atom, or stated another way, the isotope's atomic mass. Carbon's most common isotope has six protons and six neutrons; hence, it is called ^{12}C. The rarer stable isotope, ^{13}C, has an extra neutron, making it slightly heavier.[2] Generally, stable isotopes of an element all have the same chemical properties and reactions, and they perform more or less the same functions in biological processes. Even so, the heavier isotopes will often be discriminated against in biochemical processes—such as photosynthesis in plants—simply because they are heavier and therefore take more energy to process, making them, in a sense, more costly for a plant.

In this chapter we will see how the ratios of carbon isotopes can be measured in modern plants and animals as well as in ancient plant, animal, and human remains recovered from archaeological sites, giving us a glimpse of the diets that our long-deceased ancestors might have had. Although we are particularly concerned with tracking the changing role of maize in ancient diets, as we will see, there are many other isotopes besides ^{12}C and ^{13}C that can be studied—especially those of nitrogen—and they each tell us something different about the diets of ancient peoples and can even give us clues about the environments where they lived. First let's look at how this process works with carbon isotopes.

STABLE CARBON ISOTOPES AND PLANT METABOLISM

Most of us don't realize that when we eat maize, or anything made out of maize—such as corn tortilla chips, soda drinks made with high fructose corn syrup, and beef raised on maize—we are loading our bodies with more of a slightly heavier "variety," or **isotope,** of carbon than we would normally take in if we were eating other common foods made from wheat, rice, or potatoes. We don't notice because this heavier carbon isotope, ^{13}C, acts the same (and tastes the same—that is, it is completely tasteless) as the much more common ^{12}C. As a result of the particular way that maize and other tropical grasses in the Poaceae family metabolize carbon, they absorb slightly more of the rare heavier ^{13}C isotope, relative to ^{12}C, than is taken up during photosynthesis by the majority of species in the plant kingdom.

The more common type of plant metabolism—whereby plants take in nutrients and convert them into energy using photosynthesis, also called **fixation**—is called the **C3 cycle** or **Calvin-Benson cycle,** discovered in the 1940s.[3] About 99 percent of plant species on earth (and about 95 percent of all the plant biomass) use this metabolism process—which is thought to be the ancestral type. In the 1960s biochemists discovered another kind of plant metabolism, the **C4 cycle** or **Hatch-Slack cycle,** wherein carbon is processed slightly differently in hot, arid environments in order to conserve water during **transpiration**.[4] C4 cycle plants include many tropical grasses—for example, sugarcane, millet, sorghum, and some amaranths. Compared with C3 cycle plants, C4 cycle plants have a slightly higher ratio of ^{13}C to ^{12}C. Another group of plants—cacti, succulents, and their cousins—have, like C4 plants, evolved yet another metabolic process to survive and thrive in hot, arid conditions. They are collectively known as crassulacean acid metabolism plants—and their method of metabolism is called the **CAM cycle**. Like C4 plants they have, under certain conditions, higher ratios of ^{13}C to ^{12}C.[5]

In all three carbon cycles—C3, C4, and CAM—plants take in the proportions of the stable carbon isotopes that occur in their environments. But because of differences in the biochemical pathways for each cycle, they retain different proportions of these isotopes. In C3 cycle plants, the heavier ^{13}C isotope is more rapidly depleted during the carbon fixation process. C4 cycle plants retain more of the ^{13}C carbon that they take in, as do many CAM plants. As we will see below, when stable carbon ratios are measured using mass spectrometry, C3 and C4 plants can be distinguished from one another because the ratios of ^{13}C to ^{12}C (usually expressed as $\delta^{13}C$ ‰) are distinctly different. C3 cycle plants have $\delta^{13}C$ values in the -29 to -25 ‰ range, while C4 cycle plants have $\delta^{13}C$ values in the -16 to -10 ‰ range.[6] The nonoverlapping and distinct difference between these ratios extends to the consumers of these plants. People who eat strictly C3 cycle plants have body tissues that closely mirror the stable carbon values of those plants, and, likewise, people who eat only C4 plants have stable carbon values reflecting those plants.[7] Of course this is not an exact one-to-one relationship, but the patterns are so distinctly distinguishable that the ratios can be used to determine the relative contributions of C3 and C4 plants to people's diets by analyzing their body tissues.

In the late 1970s two South African paleoecologists, John C. Vogel and Nikolaas J. van der Merwe, published the first papers exploring the implications of these carbon cycle variations for tracking the appearance of maize in the diet of ancient North Americans, and as a result they helped launch an increasingly important area of archaeological research: archaeo-isotope studies.[8] Their research question was relatively straightforward: is it possible to use stable carbon ratios to find out when maize agriculture appeared in northeastern North America? The northeastern United States was an ideal place to test this method because very few naturally occurring C4 plants (which are more common to the tropics) are found in the temperate northerly latitudes, where C3 plants predominate. In such environments, ancient peoples should have had relatively low $\delta^{13}C$ values (in the -20 ‰ range). Since maize evolved from teosinte, a C4 grass native to the seasonally arid tropics, it has a very high $\delta^{13}C$ value (ranging from about -12 to -8 ‰, with an average of -10 ‰), and people with a significant portion of maize in their yearly diets would be expected to have elevated $\delta^{13}C$ values—maybe not as high as those of 100 percent corn eaters, but somewhere approaching -12 ‰. Before seeing how van der Merwe and Vogel tested their question, let's look briefly at how these isotope values are measured.

MEASURING STABLE ISOTOPES
USING MASS SPECTROMETRY

Beginning almost a century ago physicists and analytical chemists working on both sides of the Atlantic developed modern mass spectrometers—highly specialized instruments able to sort atoms by their mass, or atomic weight.[9] Using these instruments—which have become much smaller and more complex over the decades since they were first invented (they are now about the size of an apartment refrigerator)—scientists can estimate how much of a given isotope is present in extremely tiny samples of organic and inorganic substances (figure 7.1). In the case of our examination of the ancient use of maize, samples of maize and the people and other creatures who may have eaten it can be tested using mass spectrometry to determine the ratio of ^{13}C to ^{12}C in their tissues.

To test ancient human skeletal remains, analysts take a small sample of bone, usually only a gram or two, and process it to extract the proteins and minerals it contains. **Collagen** is the protein component of bone, and, like all parts of the body, it preserves a record of the elements that went into building it—such as carbon, nitrogen, and oxygen—all of which ultimately come from the person's original diet over the course of months and years. The mineral component of bone and teeth is **apatite**—the portion of the skeleton that helps to make it rigid. Of the two components, collagen is the portion that is most commonly analyzed, since it is thought to be less affected by postmortem contamination, which can be caused by factors such as water percolating through the skeletal remains during centuries or millennia of being buried in the ground.

Once extracted from the bone, the sample of collagen or apatite is put into a small capsule and placed within the combustion chamber of

FIGURE 7.1. Postdoctoral fellow Paul Szpak prepares a bone collagen sample for mass spectrometer analysis at the University of British Columbia Archaeology Isotope Lab. (Photograph by Michael Blake)

a mass spectrometer. When heated to a very high temperature, the sample releases carbon dioxide (CO_2), which is then ionized (electrically charged) and passed by a magnet en route to a detector. Isotopes with a heavier atomic weight (or mass) are deflected less by the magnet than are the lighter isotopes, causing them to be sorted by weight, thereby allowing the detector inside the mass spectrometer to calculate the relative quantities of each. So, for example, a collagen sample taken from a person who has eaten a great deal of food derived from $C4$ plants will have collagen with a high proportion of ^{13}C relative to ^{12}C, and this increased frequency of ^{13}C isotopes will show up in the spectrogram produced by the mass spectrometer's detector.

Carbon is not unique in having stable isotopes of differing atomic weights that can be detected in mass spectrometers. All chemical elements come in different isotopes that vary in atomic weight. While most naturally occurring isotopes are stable and will last an eternity, some are unstable and decay, giving off radioactive energy over time until they reach a more stable version of the original element or another, closely related one. We will return to this matter later, but first let's look at the results of Vogel and van der Merwe's very first stable isotope study of the ancient peoples of the northeastern United States.

STABLE CARBON EVIDENCE FOR THE ADOPTION OF MAIZE AGRICULTURE IN NORTHEASTERN NORTH AMERICA

Vogel and van der Merwe reasoned that in parts of North America, such as New York State, where maize was a relatively late introduction to the economy (at the time of their work in the 1970s maize was thought to have appeared sometime after about AD 1000) and where there was no other source of ^{13}C-enriched food, it should be possible to see a significant difference in δ^{13}C values between earlier non-maize eating peoples and later maize-eating peoples. In other words, the pre-maize populations should have lower δ^{13}C values than the later maize growers. To test this hypothesis they examined skeletal remains collected by archaeologists from two non-maize sites—Frontenac Island, dating between 4500 and 4000 BP, and Vine Valley, dating between about 2400 and 2000 BP—and two later sites where, presumably, people had been eating maize—the Snell site, dating from 700 to 1000 BP, and Engelbert, dating about 400 to 500 BP.

They extracted collagen from small samples of bone from three early-period individuals and four later-period individuals and found that the early people had δ^{13}C values averaging -19.7 ‰, while the later people had δ^{13}C values averaging -14.4 ‰. For the earlier people to have values near -20 ‰, they must have been eating only C3 plants or animals that ate C3 plants. Vogel and van der Merwe estimated that for the later people to have δ^{13}C values around -14 ‰, somewhere between 24 and 47 percent of the carbon entering their bone collagen must have come from C4 plants. Since maize was the only locally available candidate for a dietarily important C4 plant food, they were able to demonstrate that carbon isotope analysis was a powerful tool for helping reconstruct past diets and, in particular, the introduction of maize into the region. Since the publication of this work, there have been hundreds of studies using stable isotope analysis to illuminate the paleodiet—and a significant number of these have been aimed at tracing the changing role of maize agriculture in the Americas. Before looking at some more recent examples of isotope studies that focus on maize as part of the diet, we need to take a look at another important element—nitrogen—because it too can help us understand ancient diet.

COMBINING NITROGEN AND CARBON STABLE ISOTOPES

As with carbon, nitrogen is an important element in all plant life, and therefore all life on Earth. Nitrogen's most common isotope, ^{14}N (comprising 99.636 percent of all nitrogen in nature), has seven protons and seven neutrons, while the heavier and much more rare ^{15}N isotope has seven protons and eight neutrons (comprising only 0.364 percent of all nitrogen). Nitrogen makes up about 78 percent of the earth's atmosphere, occurring mainly in the form of a clear gas, N_2, although it is also found in the form of various, more complex chemical compounds in soil and living organisms. The two stable isotopes of nitrogen work equally well in the biological processes that require them, but, as with the heavier of the two carbon isotopes, the extra weight of ^{15}N means that it is processed differently at the cellular level than ^{14}N, leading it to be concentrated in higher proportions the farther one moves up the food chain. As a result, the protein of top predators will have higher $\delta^{15}N$ values (i.e., higher ^{15}N to ^{14}N ratios) than the protein values in the foods they consume—on the order of 2 to 5 ‰.[10]

This trophic-level difference in nitrogen isotope ratios makes it possible to use $\delta^{15}N$ measures, along with $\delta^{13}C$ measures, to make finer distinctions among diets than would be possible by using only one of the values. Over the past several decades researchers have compiled thousands of measurements of both $\delta^{15}N$ and $\delta^{13}C$ in plants and animals living in different environments and at different trophic levels in the food chain. This has allowed paleodiet analysts to create the graph that we see in figure 7.2, which shows the stable carbon and nitrogen values for broad groupings of plants and animals. Just as we saw earlier—that there is a clear distinction between the $\delta^{13}C$ values of C3 and C4 plants—here we see that the same is roughly true of the $\delta^{15}N$ values of C3 and C4 plants and the animals that consume them. However, figure 7.2 also shows that C4 plants have about the same high $\delta^{13}C$ values as

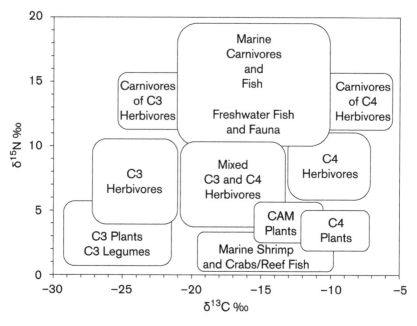

FIGURE 7.2. Baseline chart showing the distribution of stable carbon and nitrogen values for different organisms in various environments. (Illustration by Michael Blake)

some categories of aquatic animal species, including some varieties of reef fish, other marine fish, and some freshwater fish. Therefore we wouldn't be able to distinguish between them using $\delta^{13}C$ values alone, but by adding $\delta^{15}N$ measurements, we can. This means that it is possible, using measures of both stable nitrogen ratios and stable carbon ratios, to distinguish whether a high $\delta^{13}C$ value is a result of a diet high in C4 plants or high in aquatic foods. This differential has proven quite useful in cases where we want to know how significant maize may have been in the diets of coastal or riverine dwellers, where maize and other C4 plants don't grow naturally but may have been introduced at some point in the past.

In the same way, it is possible to distinguish whether individuals have very low $\delta^{13}C$ values as a result of eating many C3 plants or eating large quantities of C3-eating animals. Those who consume primarily C3 plants will have lower $\delta^{15}N$ values than those who eat C3-plant-eating herbivores. This distinction between nitrogen and carbon levels makes it possible for us to see significant changes in diet, particularly in regions where there can be mixed diets with food coming from multiple sources.

MAIZE EATERS IN MESOAMERICA

Throughout Mesoamerica many hundreds of archaeological sites have been found to contain the well-preserved remains of the people who built and occupied them centuries and millennia ago. In many cases their descendants continue to live in the area today and practice many of the same types of agriculture as their ancestors. We know, from histories of the region and from archaeological evidence, that maize was and still is an exceedingly important part of the Mesoamerican diet. Stable isotope analysis of human bones—and in some cases animal bones—has confirmed this very clearly. But going beyond the simple demonstration of whether or not people consume maize, this type of analysis has also made it possible to show that there are sometimes clear differences in the amounts of maize that were consumed by people in different communities, and even by different members of a single community. These differences may reflect local environmental variation, social differences, or food preferences. Let's look at a few examples that illustrate the dietary distinctions that have been inferred based on stable isotope analysis.

The bio-archaeologist Christine White and her colleagues have studied a number of samples from Maya sites located in Belize. At the site of Pacbitun, for example, they tested thirty-three burials, twenty-one of which were well enough preserved to yield reliable results. Most of the samples they recovered dated to the Terminal Classic period (about 1200 to 1000 BP), although some were older.[11] Stable carbon and nitrogen isotope analysis shows that most individuals dined mainly on C4 plants—presumably mostly maize, but there were important differences in maize consumption depending on status, residential location, age, and gender. White and her team found that the individuals buried in the most elaborate tombs and with the most grave goods—and therefore considered to have been high status—also had the highest $\delta^{13}C$ values. Individuals in simpler graves who had lived farther from the ceremonial center had lower $\delta^{13}C$ values. They also found that on average males consumed more maize than females, and adults more than juveniles.

White and her colleagues have also analyzed the stable isotopes in human bones at two other sites in Belize: Lamanai and Altun Ha, both excavated by David Pendergast of the Royal Ontario Museum.[12] These sites are farther north in Belize than Pacbitun and are situated closer to the coast. At both sites a number of higher status people, some of them probably members of the ruling lineage, received a significant amount of their nutrition from seafood—probably fish from the highly productive marine/reef ecosystem located just offshore. Analysis of the bone collagen and apatite of two adults—a female and a male—found in the most elaborate tomb at the Lamanai site revealed that the female had stable carbon and nitrogen isotope ratios that could only have resulted from a diet very high in C_4 plants—almost certainly maize—higher in fact than anyone else at the site. The adult male, in contrast, had carbon and nitrogen isotope values that, considering the local environment, were most likely the result of a much higher consumption of both seafood and "maize-fed terrestrial animals" than anyone else at the site.[13] At Altun Ha, especially in the earlier periods, a significant proportion of the population ate a combination of C_4 and C_3 plants, along with aquatic foods and terrestrial animals that were part of a C_3 nutrient system. But in later periods the nutritional patterns became much more complex. By the Late to Terminal Classic periods maize consumption decreased overall, a trend that continued into the Postclassic period a few centuries later. Even so, social distinctions persisted. Low-status people living on the outskirts of the city were the most reliant on plant foods (and, increasingly, plants other than maize), while some elites living in the center consumed a diverse range of foods including maize, seafood, and terrestrial meat and other elites seemed to have had a distinct preference for seafood.[14]

Studies such as these alert us to the fact that the patterns of food consumption in the past, just as in the present, could be highly variable within and between communities and over long periods of time. As archaeologists dealing with very small samples, highly variable preservation conditions, and, at best, a partial glimpse of what people in the past ate throughout the months and years of their lives, it is not surprising that we often reduce our findings to a very simplistic model. In the case of the sites we just looked at, plant preservation was fairly poor, and animal bone preservation was only marginally better. The ancient food remains recovered at the sites likely represent only a small portion of the full range of foods people ingested. The stable isotope analyses that have been carried out help us see several more dimensions of the ancient food economy and individual nutritional patterns than we would be able to if we simply studied the food remains that happened to have preserved in the trash deposits of these long-abandoned villages and towns.

As we have just learned, high $\delta^{13}C$ values can be produced by the "marine effect." Since marine foods produce $\delta^{13}C$ values that are similar to those produced by C4 and some CAM plants, we have to rely on $\delta^{15}N$ values to tell which food sources most likely contributed to the elevated $\delta^{13}C$ values. We face a comparable dilemma in other parts of Mesoamerica because other plants besides maize can raise $\delta^{13}C$ values. In the arid tropical regions of the Americas there are several species of plants that have $\delta^{13}C$ values very similar to maize. These include millets, amaranth, and some CAM plants—such as agave and prickly pear—plants that we know people were eating at certain times and in certain regions in significant quantities.[15] This means that people who were consuming such foods would have had elevated stable carbon values that were not solely the result of eating maize. It also means that while stable isotope analysis can help confirm the possibility that a food such as maize was a significant contributor to the diet, it doesn't rule out the possibility that other foods that produce similar stable isotope ratios might have also been consumed. One would need to look at other lines of evidence—for example, macrobotanical and microbotanical remains—to be certain of the major dietary contributors.

This isn't so much of a problem in the later periods of Mesoamerican history because so many lines of evidence converge to tell us not only that maize was a major part of the diet but also that it was the most important staple and that it was both symbolically and ritually at the center of most Mesoamerican cultures. But when we move back in time to the periods when maize was newly domesticated, or when it was newly introduced into a region, its dietary significance can be much more difficult to discern. In some regions, such as eastern North America, this wasn't an issue because it was relatively easy to distinguish between early non-maize eaters and later maize eaters, since there were essentially no dietary sources other than maize that could have significantly elevated people's $\delta^{13}C$ values. The situation is quite different, however, in tropical regions in the Americas.

One example of this slightly more confusing picture comes from the Soconusco region of southeastern Mesoamerica, where, since the late 1980s, my colleagues and students and I have been studying the question of when maize became an important part of the ancient diet. As in many parts of Mesoamerica that are outside teosinte's homeland in west-central Mexico, there are several sources of food in the Soconusco region that could produce fairly high stable carbon ratios. In our excavations we have recovered human bone samples from archaeological

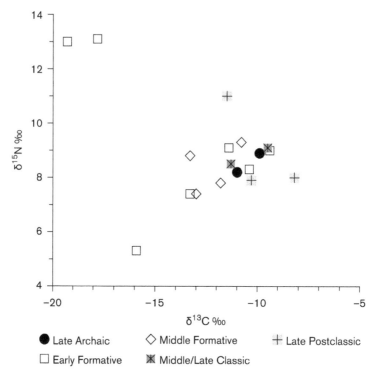

FIGURE 7.3. Stable carbon and nitrogen measurements for the archaeological sample of humans in the Soconusco region of Pacific coastal Chiapas, Mexico. (Illustration by Michael Blake, based on data presented in Moreiras 2013)

sites spanning the past 5,000 years. Unfortunately, many of them are not well preserved, and it has proven difficult to extract enough uncontaminated organic material from them in order to perform stable isotope analysis. However, we managed to extract collagen samples useful for testing from approximately nineteen individuals. We also tested samples from another thirty individuals, but the bone turned out to have been too badly decayed or contaminated from the percolation of groundwater through the soil, so the results were not meaningful.[16] Figure 7.3 shows the distribution of the samples through time—from the Late Archaic to the Postclassic periods. One thing that is apparent is that there is a great deal of variation. Some samples look to be heavily influenced by a more aquatic diet (either riverine or marine), others appear to have been heavily C_3 based, and others seem to tend toward a C_4 or CAM plant-based diet. These findings are not surprising, because the village and town sites in the region are located adjacent to the Pacific Ocean, next to vast fresh and brackish water estuaries, swamps, and freshwater rivers that flow into the ocean. This is similar to other parts of lowland Mesoamerica, where, although people have been practicing plant cultivation for at least 6,000 years, they simultaneously hunt, fish, and gather wild plants and animals. Like the ancient Maya people we looked at in Belize, who lived in settlements close to the coast, the inhabitants of the Soconusco region had a rich array of aquatic resources to choose from, many of which were, and still are, available year-round.

Returning to the pattern seen in figure 7.3, it appears that there was a heavy reliance on C4 plants—possibly maize—in the late Archaic period (approximately 4,000 to 5,000 years ago). The two individuals from the site of Tlacuachero, excavated by Barbara Voorhies in the early 1970s, had $\delta^{13}C$ values and $\delta^{15}N$ values that are most similar to what we would expect from maize (or other C4 plant) consumption. These people lived for part of the year in the vast estuary system on the northern edge of the Soconusco and, Voorhies argues, moved inland for the rest of the year, presumably to hunt, gather, and tend crops. Unfortunately, no well-preserved late Archaic period sites have been found in the inland zones, so we don't have any examples of these people's dwellings, trash deposits, or other materials to examine. Still, it is puzzling that their stable carbon and isotope ratios are closer to what we would expect from—and actually see among—the later period maize farmers living alongside the productive estuary system.

In an attempt to understand the range of variation in stable isotope values in the base population of plants and animals on which the people living in the Soconusco region would have subsisted, I sampled modern-day plants and animals in the area with Brian Chisholm, an archaeologist specializing in isotope analyses and a colleague of mine at the University of British Columbia. We found that there were a number of species of estuary shrimps and crabs that mimicked the stable isotope values we might expect from C4 plants. If we consider these aquatic resources as a potentially significant component of the late Archaic period diet, it might explain why the isotopic values of these people looked like those of C4 (maize) or CAM plant consumers.[17]

By the Early Formative period, people began to rely more heavily on agriculture, including the cultivation of maize. The stable isotope values clearly show that maize was becoming an increasingly significant part of the diet for some people. However, during this period, when the first settled villages started to show up in the region, it also appears that there was a great deal of variation, similar to what we observed at the sites of Pacbitun, Lamanai, and Altun in Belize. Some individuals appear to have been eating very little maize, while others may have consumed as much as present-day farmers in the region. This variation continued to a certain extent through the Middle Formative period—although we see that the reliance on C4 plants was becoming clearer, and none of the individuals tested showed the very low stable carbon values that we found in the previous period. We don't have any examples from the Early Classic period in this region, but by the Middle and Late Classic, the people who lived close to the estuary look to have had significant C4 diets, and maize is the most likely source. The evidence shows that by about 3,000 years ago, maize had become a dietary staple for many people in the Soconusco region, and by 1,500 to 2,000 years ago, everyone was dependent on it.

MAIZE CONSUMPTION ALONG THE ANDEAN COAST OF SOUTH AMERICA

Thanks to the extremely cold Humboldt Current that flows northward along the Pacific coast of northern Chile and Peru, there is so little rainfall that organic remains at open-air sites preserve almost as well as those in dry cave sites. This means there are many more extremely well-preserved plant, animal, and human remains recovered in archaeological sites in this area than in almost any other region of the Americas. This makes the area ideal for studying the stable isotope values of large numbers of ancient people. Not surprisingly, one of the key observations from these studies is that people had diets high in marine foods—this coast is the epicenter of the world's most productive fishery and has been for many thousands of years.[18] A large-scale stable carbon and nitrogen isotope study of fifty individuals from eight village sites in the Viru Valley of Peru, spanning the period from about 2500 to 900 BP, shows that, while they had access to maize by at least the Early Horizon, it was probably an ancillary part of the diet and not a major staple for most people.[19] Jonathon Ericson and his colleagues, who carried out these analyses, thought that some of the high $\delta^{13}C$ values were likely a result of marine food consumption. It wasn't until the Gallinazo culture, which began by about 1800 BP, that there was evidence of a significant increase in maize consumption, but even then it was part of a gradual trend. The team suggests that the increase might have been the result of new, more productive hybrids of maize coming available.

These finding are similar to those of Richard Burger and Nikolaas van der Merwe, who examined the stable carbon isotope values of human remains from the one of Peru's most important ancient sites—Chavín de Huantar, which is located in the highlands of northern Peru and is the type site for the Early Horizon period—as well as two samples from the nearby site of Huaricoto. The $\delta^{13}C$ measurements they obtained from samples dating from about 2600 to 2300 BP were very low (ca. -18 ‰)—indicating that, although maize was present, it was not an important staple food. Burger and van der Merwe suggest an interesting alternative interpretation: instead of consuming maize as a solid food, people may have fermented and brewed it to make the maize beer known as *chicha*—a beverage that was highly esteemed in many regions of the Andes, both in the past and right up to the present day.[20]

In another, more recent isotope study of two Early Horizon (ca. 2500 to 1800 BP) sites and one Early Intermediate period site in Peru (Pacopampa, located in the northern highlands, and Mina Perdida and Tablada de Lurín along the central coast), Robert Tykot, along with Burger and van der Merwe, found a pattern similar to their earlier study.[21] Measurements from the collagen component of ten rib samples from Pacopampa gave very consistent $\delta^{13}C$ results, yielding an average of −19.3 ‰. They also measured the carbonate—apatite—portion of the bone and found slightly more evidence for some maize, or C4, consumption—the average for the samples was -9 ‰.[22] This could have been the result of consuming maize as *chicha* beer rather than solid food, since the *chicha* would elevate the $\delta^{13}C$ values of the apatite and not the collagen—a topic we will discuss in more detail in the final section of this chapter. At Mina Perdida the team tested hair samples from nine people (that's how good the preservation is on the coast!) and found an average $\delta^{13}C$ of -18.1 ‰, along with a relatively high average $\delta^{15}N$ value of 10.6‰, which is highly suggestive of a reliance on marine foods. They took only five samples from the youngest of the three sites, Tablada de Lurín, which gave an average $\delta^{13}C$ value of -11.2 ‰ and an average $\delta^{15}N$ value of 14.4 ‰. These results seem to indicate that there was likely a very high marine food component to the diet—but because the team also measured the $\delta^{13}C$ values of the apatite in these samples, they could see that the high apatite measures, with an average of -5.7 ‰, suggest a high consumption of maize (or other C4 plants).

In comparing diet histories in parts of Mesoamerica with those in the Andean region, even if just looking at a few examples, we can see some fascinating similarities and differences in the timing and the degree of significance of maize cultivation and consumption among these peoples. We know that maize farmers in Mesoamerica were drawing on very long local traditions of maize growing. So when we see an increase in $\delta^{13}C$ values from the region through time, it's likely that it reflects changing proportions of maize use—even during the Archaic period (pre-4000 BP), before we have much evidence of permanent villages. By comparison, maize was introduced somewhat later to the Andean region and was not part of the original botanical repertoire there. But when it did arrive along the Pacific coast and adjacent Andean highlands of South America, it was coming into what were already sedentary communities with either complex agricultural economies or highly developed and productive maritime economies. In both cases maize had to be integrated into the local economies in new ways, either competing with or integrating with an already complex array of foods that had long histories in the area. We know that by the time maize reached Andean South America, people already had a wide range of staple crops that accomplished what maize was able to do for early Mesoamericans—that is, ensure a reliable and storable supply of carbohydrates. They had long grown an enormous range of productive and nutritious food plants, including the potato (*Solanum indigenum*), sweet potato (*Ipomoea batatas*), ullucu (*Ullucus tuberosus*), oca (*Oxalis tuberosa*), acira (*Canna edulis*), manioc/cassava (*Manihot esculenta*), quinoa (*Chenopodium quinoa*), beans (*Phaseolus vulgaris*), tomato (*Solanum lycopersicum*), and lucumo (*Pouteria obovata*), to name a few. For early Central and South Americans, maize was very likely simply one of many useful plants when it first arrived—and perhaps even for several millennia. I find it very plausible that they initially found maize to be useful in the production of beer—*chicha*—yet, again, this is something that they had perhaps been doing very successfully with a wide range of other high sugar or carbohydrate plants, such as manioc. It might have been that maize was convenient for this purpose because it grew in conditions and environments where some of the traditional sources of carbohydrates may not have flourished; for example, some of the root crops, like potatoes, were native to the cool highlands, but they didn't do so well down on the arid coastal lowlands. Maize may have helped people to produce storable carbohydrates—both for eating and for making beer—in regions where there were few other such alternatives.

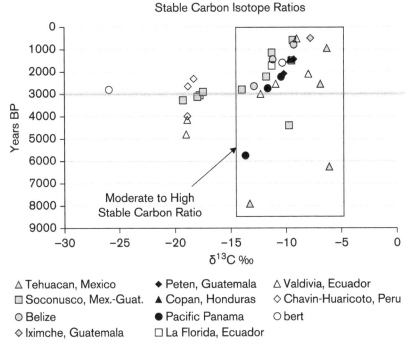

FIGURE 7.4. Graph showing chronological changes in stable carbon isotope values across the Americas. It represents the published measurements for over six hundred individuals and shows a significant increase in stable carbon isotope ratios after about 3,000 years ago, likely corresponding to peoples' increasing reliance on maize. (Illustration by Michael Blake, after Smalley and Blake 2003)

Stable isotope analyses of ancient Americans' diets shows that the local importance of maize varied through time, from community to community, and often between different social groups within a given community. Still, broad changes in the role of maize across continents can be observed by looking at the timing of the shift from low to high use. By combining the results of dozens of isotope studies carried out in recent years, it is possible to glimpse these broad changes. At the earliest end of the sequence of *Zea mays* use, we expect there to be little maize-induced rise in $\delta^{13}C$ values. In regions where we know maize was not present, but where $\delta^{13}C$ values measured in human remains are nonetheless high, we must conclude that some other C4 plant—or perhaps CAM plants or particular aquatic species, such as fish—is responsible for the high stable carbon values. At the late end of a historical sequence, in a region where we know that maize had been in use for centuries or millennia, we expect to see very high $\delta^{13}C$ values, reflecting the increased reliance on maize as a food crop—such as those we saw for the Maya and Soconusco regions of Mesoamerica.

Figure 7.4 shows the results of a study I conducted a few years ago as part of the work that John Smalley and I did when hypothesizing that one of the initial motivations for the early movement and adoption of maize was to harvest a sugar from the plant's stalks—a source of sweet juice that could be fermented to make alcohol.[23] In this graph we show that, based on over six hundred published stable carbon measurements that were available in 2003, there was an overall gradual trend toward increasingly high $\delta^{13}C$ values, reaching a threshold of what we term "moderate to high" stable carbon ratios by about 3,000 years ago, when they are on average higher than -15 ‰. In this graph many of the samples postdating 2000 to 1500 BP show very high values: greater than -10 ‰. The few anomalies—that is, earlier samples that are also very high—are from individuals that were very likely eating either marine foods or CAM or other C4 plants. The majority of the post-3000 BP people represented in this sample were probably maize eaters.

FOOD VERSUS DRINK: USING ISOTOPES TO DISTINGUISH BETWEEN THE TWO MODES OF CONSUMPTION

Let's return to a question I first raised in the introduction: can we distinguish, isotopically, between maize eaters and maize drinkers? Over a decade ago, as we were working on the Soconusco isotope samples, my colleague Brian Chisholm suggested that the metabolic pathway of carbon could have a significant impact on whether maize would be visible isotopically in the collagen versus the apatite. We reasoned that it is well known that the protein we eat is used by the body to produce the proteins that make up our bone collagen—along with many other proteins of which were are composed. The carbon in the carbohydrates and fats (lipids) that we eat does not contribute much to our proteins, with the possible exception of extreme protein starvation.[24] This is different from the carbon that makes up the apatite portion of our bones, which comes from a broad mix of all three macronutrients that we consume—proteins, carbohydrates, and fats, the majority of which, however, is derived from carbohydrates.[25] Therefore, because alcohol is a carbohydrate, carbon that comes from the ingestion of alcohol should be primarily visible in the stable carbon isotopic signature measured in bone apatite and relatively invisible in bone collagen. Unfortunately there have been few experiments to demonstrate these carbon pathways.

Because we couldn't do this experiment on ourselves, we used rats as a proxy for humans to see if there were clear differences in the stable carbon ratios between maize beer drinkers, maize eaters, and (non-maize) C3-food eaters. Cecilia Canal, our graduate student, who conducted this research for her master's thesis project, showed that, at least among the rats who dedicated their lives to this experiment, there are clear and significant differences between the three groups. Her main finding was that the C4 maize-based carbon in the maize beer did indeed register in the bone apatite of the rats but not in their collagen. This is exactly what we expected if the majority of carbon in collagen comes from ingested protein—something pretty much lacking in the maize beer diet—while carbon in the rats' apatite came from the mix of carbon from carbohydrates in the alcohol and fats and proteins from the non-alcohol portion of their diet.[26]

One important implication of this study is that, if early teosinte and/or maize users had been making beer from the stalks of these plants—as Smalley and I, along with Hugh Iltis, proposed—we shouldn't see an isotopic signature of *Zea mays* consumption, even though people might have been deriving a significant portion of their calories from this beer. This has even more profound implications for the isotopic visibility of *chicha* beer drinking among early Central and South American maize farmers. It is quite possible that the stable carbon isotope ratios as measured in bone collagen would show little or no consumption of maize if people were consuming most or all of their maize in the form of *chicha*. On the other hand, the stable carbon isotope values of their bone apatite should be quite high—perhaps just as high as if they were eating maize—which is exactly what we saw in our population of rodents that enjoyed a diet heavy with maize stalk beer. In chapter 9 we will take a closer look at the social and symbolic significance of maize beer and at how important it was and still is in Andean society.

CHAPTER 8

Genetically Modified Maize the Old Way—By Agriculture

On the molecular level, new genetic research is bringing into focus the changes teosinte's genome underwent as it evolved under domestication. This is a complex and specialized area of research—and one that falls squarely within the realm of plant genetics. But because geneticists, archaeologists, and paleoethnobotanists are interested in many of the same questions, it is imperative that we understand each other and the implications and potential pitfalls of each other's work. This chapter introduces some of the new genetic discoveries in maize research—with as little jargon as possible and with explanations of the necessary terms and techniques. Maize has one of the best-studied **genomes** of all domesticated plants, and almost every month geneticists report significant new discoveries about *Zea* genetics. These can be overwhelming for the nonspecialist, but it is becoming essential to understand this branch of science because it transforms our understanding of all aspects of life, including the plants and animals that our ancestors domesticated.

In this chapter I summarize two main lines of genetic research, one much more advanced than the other: studies of modern maize and its relatives and studies of ancient maize **DNA.** The first looks at the DNA of present-day members of the genus *Zea* to discover, compare, and interpret the genetic variations among all its species and subspecies. Of great interest is the search for "bottlenecks," periods during which genetic variation in ancestral species was reduced as ancient peoples

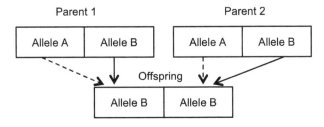

FIGURE 8.1. General pattern of allele inheritance, showing that each parent contributes one of their two copies of a particular gene to their offspring. (Illustration by Michael Blake)

selected maize plants with particular characteristics, such as single rather than multiple stalks, larger cobs with more rows and larger kernels, drought resistance, increased starch content, increased sweetness, and so on.[1] Geneticists are increasingly able to pinpoint the exact **genes** and their variants, or **alleles,** that are responsible for traits that were of interest to the first maize farmers. The allele variants that are passed from parents to offspring determine the genotype of each new generation, and the **genotype,** in turn, determines the **phenotype**— or the genetic expression of any given set of traits that are inherited (figure 8.1).

Another promising, but more difficult, avenue of genetic research is the study of ancient DNA recovered from archaeological samples. This type of DNA analysis is usually abbreviated as aDNA. It is more challenging because ancient DNA is so fragmented—yet new gene cloning technology is making it possible to find enough intact genetic material to compare the alleles of maize samples from different regions and different time periods. In one of the first successful examples of ancient maize DNA analysis, Vivian Jaenicke-Després, Bruce Smith, and their colleagues were able to extract DNA from pieces of maize that were as much as 4,400 years old, both from Mexico and the US Southwest.[2] They found evidence that some low frequency teosinte alleles that control aspects of plant architecture (*tb1*), starch production (*su1*), and protein production (*pbf*) became increasingly common (and eventually fixed) in maize as the desirable variants of these genes were selected because of their preferred characteristics. As more specimens are tested it will be possible to track many more such genetic changes, both through time and across the continents.[3]

IDENTIFYING TEOSINTE AS MAIZE'S ANCESTOR

In 2002 a five-page paper appeared in the *Proceedings of the National Academy of Sciences, USA* that seemed to put to rest any lingering doubts about the ancestry of maize. A team of botanists working with John Doebley in his evolutionary genetics lab at the University of Wisconsin, Madison, presented the results of a painstaking genetic study of maize and its relatives that had as its goal answering the question, what was the ancestor of maize?[4] This simple query had eluded botanists and archaeologists alike for many generations, partly because of the great diversity of modern maize (which raises another question, how do we account for such diversity in a single species with just one ancestor?) and partly because of the great geographic spread of both maize and its putative ancestors—which included various species of teosinte, *Tripsacum,* and the elusive wild maize. Research published in 1984 by Doebley and his colleagues, who looked at **isozymes** (structurally similar variants of enzymes) in order to compare allele variations between maize and all the teosintes, confirmed that the isozymes of *parviglumis* were virtually "indistinguishable from those of maize."[5]

Doebley and his team's 2002 study, with Yoshihiro Matsuoka—who was at the time a postdoctoral fellow in the Doebley Lab—as the lead author, demonstrated that *Zea mays* ssp. *parviglumis* was the only subspecies of teosinte with enough genetic similarity to all the varieties of modern maize to be its common ancestor. At the same time the study confirmed previous hypotheses that the Balsas Valley region of southern Mexico, where the natural range of *parviglumis* is located, must have been where the first farmers began to domesticate the plant. Here's how Matsuoka, Doebley, and their research team did it.

They began by scouring existing botanical collections for different varieties of maize and three subspecies of teosinte (*parviglumis, mexicana,* and *huehuetenangensis*), selecting 264 samples that spanned the widest possible pre-Columbian biogeographical range of the plants (from Canada to Chile and from coastal lowlands to the high Andes).[6] Their samples included 193 maize and 71 teosinte plants, from which they selected DNA material for **genotyping.** Genotyping entails taking DNA stretches that contain repetitive units of one to six **base pairs (bp),** called **microsatellites,** and selectively amplifying them using a process called **polymerase chain reaction (PCR).** The amplification produces enough DNA so that it can be studied—otherwise the samples would be too small to observe.

The small chunks of DNA that contain microsatellites are snipped out of the original DNA chain using molecules called **primers** that are tailor-made for the job. In the Matsuoka study the researchers chose ninety-nine different microsatellites widely distributed in different locations across the maize genome. The repeating base pairs in microsatellites are often distinct enough that they can be used as markers—allowing the assessment of genetic similarity among different individuals. In other words, two individuals who are genetically related are likely to have a greater degree of similarity in their microsatellite markers than genetically distant individuals. The repeating nucleotides that comprise these small chunks, or microsatellites, are the building blocks of DNA and RNA. In the case of DNA, these repeating nucleotides are molecules consisting of one of the following four chemical bases, adenine (A), guanine (G), thymine (T), and cytosine (C), along with a sugar and a phosphate molecule. PCR takes the microsatellite segments and makes many exact copies so that they can be compared from individual to individual. Because a relatively large number of microsatellites were used in their study (three to four times more than had been used in previous genetic similarity studies that compared diverse species such as humans and horses or mammals and reptiles), Matsuoka, Doebley, and their research team were able to make very precise assessments of the genetic relatedness of the 264 maize and teosinte plants.[7]

Once the DNA variants, or alleles, were described for the ninety-nine microsatellites on each plant, they were analyzed statistically to create a phylogenetic tree. The team used a statistical analysis that resulted in a plot showing the degree to which genetically similar samples—based on the proportion of shared alleles—are located adjacent to one another.[8] They then used a set of computer software packages to plot the distances between samples and group them into a tree diagram, with the branches on the tree (the phylogeny) representing the genetic proximity of one sample to another. The data were sampled many times, and many trees (one thousand of them) were produced to check that the shape was not dependent on a particular combination—a process called "bootstrapping." The phylogenetic tree in figure 8.2 is the result of many such runs and is in effect an "average" tree showing the relationship of their samples, coded by their geographic origin, and plotting the connections between the many maize varieties as well as the three subspecies of teosinte.

The researchers found that a subspecies of Guatemalan teosinte, *Zea mays* ssp. *huehuetenangensis*, already known to be distantly related to the other teosinte subspecies, was the most distantly related of the samples. The next separation was between subspecies of teosinte: *parviglumis*

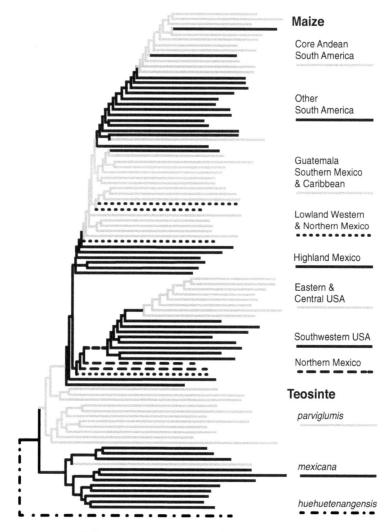

Maize

Core Andean
South America

Other
South America

Guatemala
Southern Mexico
& Caribbean

Lowland Western
& Northern Mexico

Highland Mexico

Eastern &
Central USA

Southwestern USA

Northern Mexico

Teosinte

parviglumis

mexicana

huehuetenangensis

FIGURE 8.2. Phylogenetic tree showing the relationships among nine regional varieties of maize and three subspecies of teosinte, demonstrated by the research of Matsuoka and his team (Matsuoka et al. 2002). The branching, based on genetic similarity, shows the two main movements of maize. First, the sequential spread of maize descended from the subspecies *parviglumis* northward into northern Mexico, then into the southwestern United States, and finally into the central and eastern United States; and second, the movement of maize into highland Mexico and lowland western and northern Mexico, then into the Caribbean and southward into southern Mexico and Guatemala, and eventually into South America (also see map 8.1). More recent genetic analyses have refined this picture, but the basic sequence of genetic similarity and geographic spread is the same. (Illustration redrawn by Michael Blake after Matsuoka et al. 2002:6082, figure 2b)

and *mexicana*. Finally, all of the landraces of maize branched off from *parviglumis*. This result strongly supports the case for a single origin for maize and means that it is exceedingly unlikely that any other genera (such as *Tripsacum*) or subspecies of *Zea* played a significant role in its genetic ancestry during or after domestication.

When the researchers grouped the 264 samples into 95 smaller subgroups based on their ecogeographic origin and proximity and recalculated their statistical similarity matrices, they found that the branches also corresponded to a simple geographic spread model. The population of *parviglumis* is located in the central Balsas River region and is the most likely candidate for the ancestor of all subsequent varieties of maize. The early diversification of maize seems to have taken place predominately in the highlands of central Mexico—which includes the inland portions of the states of Jalisco, Michoacán, Guerrero, and Oaxaca and the neighboring highland states of Mexico, Morelos, and Puebla—and it is there that the greatest genetic diversity is still found. The study also found that, although not a huge contributor, there is some evidence for the mixing of *mexicana* genes in maize—especially in maize growing at higher altitudes, above 1,700 meters.[9]

The genetic similarity analysis suggests that maize spread from the highlands of Mexico through western and northern Mexico and then into the southwestern United States. Simultaneously maize spread into southern Mexico, the lowlands of Central America, the Caribbean islands, lowland South America, and then into the Andean region.[10] Map 8.1 indicates the dual sequences of dispersal (northward and southward) and the potential timing as domesticated maize moved out of its homeland, where the teosintes *parviglumis* and *mexicana* are found. As we saw in chapter 4, the direct dating of maize macrobotanical remains from archaeological sites closely parallels the scenario put forth by this genetic research and confirms some of its main points. However, as Matsuoka, Doebley, and their team pointed out, there are still gaps and puzzles in our knowledge—and especially in the archaeological record of maize's migration history.

The study was also able to estimate the timing of maize's domestication, using an analysis of thirty-three microsatellites that have a known mutation rate. They estimate that *parviglumis* was domesticated sometime between 5689 and 13,093 BP (with a 95 percent probability that the true date of domestication falls within this range). The midpoint of this distribution is 9188 BP. We can narrow this date down by looking at archaeological evidence. We know that domesticated maize was present at Guilá Naquitz by 6200 cal BP, and we have not yet found any

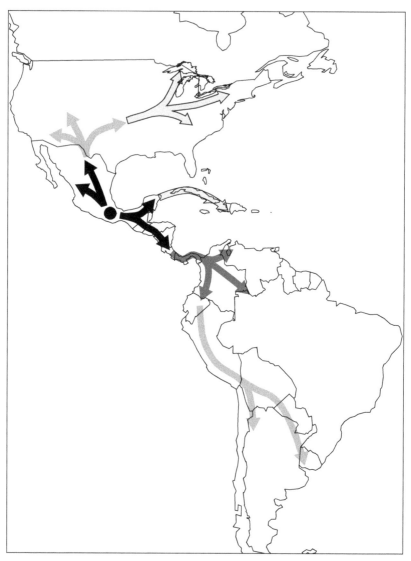

MAP 8.1. The chronological sequence of maize dispersal based on the genetic analysis presented in figure 8.2 (Matsuoka et al. 2002). The darker arrows represent the earliest spread both northward and southward. Subsequent movements are indicated by progressively lighter arrows. (By Michael Blake and Nick Waber)

archaeological remains of maize in highland Mexico that date to earlier than about 10,000 BP. One striking implication of the study's results is that maize was domesticated either in the highlands, just above 1,800 meters, or in the inland zone, between about 700 and 1,800 meters. It was not initially domesticated in the coastal lowlands of Mesoamerica or South America.

EARLY FARMERS SELECT DESIRED TRAITS

The differences between wild teosinte and domesticated maize are many, and some of them are, as botanists say, "of large effect." We will look at a few key examples of the genetic changes that teosinte underwent on its road to domestication. These changes have been discovered and described by botanists, who have been able to provide vivid descriptions of the plant's evolution. Many of the studies have come from John Doebley's group at the University of Wisconsin, but other plant geneticists working in labs in the United States and Europe have also made major discoveries. In dozens of publications during the past twenty years botanists have carefully traced the genetic differences between teosinte and maize, building a solid base of understanding.[11] Many new analyses are taking advantage of the latest gene sequencing technology. Whereas ten years ago we could hope to sequence only a handful of genetic markers, it is now possible, at a fraction of the previous cost, to sequence tens of thousands of genes. Soon we will be able to routinely sequence the entire genome of each individual plant—modern or ancient.

Still, work carried out over a decade ago has allowed a clear understanding of some of maize's most important domestication genes—genes "of large effect." Two of the better-known genes are labeled *teosinte glume architecture1* (*tga1*) and *teosinte branched1* (*tb1*), and each has profound implications for the domestication process.

Teosinte Glume Architecture1 (tga1)

Referring to Doebley's research, Hugh Iltis seized on one key mutation that must have preceded all of the other genetic changes that took place during maize's domestication process: the appearance of a mutant version of the *teosinte glume architecture1* gene—which botanists call *tga1* for short.[12] As is the case for many grasses, teosinte's seeds are protected by a complex of specialized structures, which are really just a special form of leaf. The **cupule**, a structure that supports the grain, is a set of reduced, fused, and hardened leaves called **bracts.** Another set of bracts form the **glume**—a pair of highly modified leaves that completely cover the teosinte seed. Thanks to *tga1*, teosinte's glume bracts, or hardened leaves, are almost impenetrable—which is wonderful for protecting the seed from pests during the dry season while it patiently waits to germinate, but very difficult for humans to break through when trying to harvest the nutritious grain inside.

When mature, teosinte's glumes form a stone-like cupulate fruitcase that protects the kernel, even allowing it to be eaten by animals and pass unharmed through their digestive tracts. In this way the protected grain gets a free ride and can disperse over the landscape. The mutation giving rise to the allele of *tga1* most common in maize (labeled *Tga1-maize*) reduces the size of the hard glumes so that the kernel is exposed, making it accessible to both pests and human harvesters.[13] Because humans protect the ripened, exposed grain for their own consumption and reseed the following year's crop themselves, the mutated teosinte doesn't have to be as robust as in the wild. A mutation that would lead to extinction in the wild becomes both possible and prevalent under human care.

Hugh Iltis hypothesized in vivid terms (see chapter 1) that early teosinte harvesters must have come across the exceedingly rare *Tga1-maize* mutation in a *parviglumis* plant and then collected the seeds and replanted them. I think that for this scenario to be likely, farmers must have already been planting and tending *parviglumis* outside of its natural range. This is because, within the plant's natural range, farmers would have had plenty of wild *parviglumis* seeds to gather and sow for each year's planting. But outside the natural range farmers would have to have saved seeds from their own harvest, since there would have been no wild alternatives.

A recent genetic study of maize and teosinte carried out by Huai Wang and his associates maps the precise location of the *tga1* gene and shows that it alone is responsible for the cascade of changes in teosinte that resulted in "naked grains of maize."[14] As they put it: "Maize domestication involved a change in ear development such that the cupules and glumes form the internal axis of the ear [cob], rather than casings around the kernels. In a sense, maize domestication involved turning the teosinte ear inside out"[15] (figure 8.3).

They have pinpointed *tga1*'s exact location on maize chromosome 4 and narrowed it down to a 1,042 **base pair (bp) segment** that controls the important differences in teosinte glume architecture between maize and teosinte. These differences are determined by the properties of the various alleles of *tga1*. The *tga1-teosinte1* allele, the most common

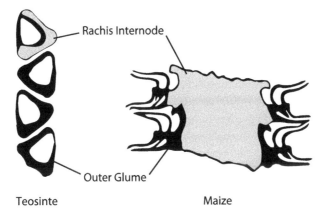

FIGURE 8.3. Illustration showing the major phenotypical changes brought about by a mutation in the *tga1* (*teosinte glume architecture1*) gene. In maize, the rachis internodes have coalesced to form the core of the cob. The outer glume—which in teosinte surrounds the kernel—retracts in maize to form an inner glume that subtends the maize kernels and provides the hard casing of the cupule. (Illustration redrawn after Doebley 2003; http://teosinte.wisc.edu/morphology.html)

form in *parviglumis,* causes the cupule to be enlarged, the glumes to completely cover the kernel, and both to be hardened with silica inside the cells (creating phytoliths) and with lignin (a durable polymer that strengthens plant cells).[16] The *Tga1-maize* allele (which Wang and his team think was a true mutant because it has not been found in teosinte) causes both the glume and the cupule to shrink and soften so that the grain is exposed and much easier to harvest. One of their most significant discoveries is that in this sequence of 1,042 base pairs, no more than 7 base pairs differ between teosinte and maize, and just one of them—an amino acid substitution, probably the lysine (K) → asparagine (N) substitution—was the crucial one. This is a very small molecular difference between these two alleles, yet one that that had a far-reaching impact on the evolution of maize—mainly because it regulated a number of other genetic expressions that depended on it. This is a topic that plant geneticists are working on, with, one hopes, "large effect."

Teosinte Branched1 (tb1)

In his review and summary of recent genetic research on maize, John Doebley outlines the remarkable insight that Hugh Iltis had when he shifted focus from the differences in ear morphology between teosinte and maize to the differences in the stem and branch morphology.[17] In an article published in *Science* in 1983, Iltis proposed that the maize ear derived from a mutation combining the male spike and female inflorescence in teosinte.[18] Unlike maize, teosinte has lateral branches containing ears (the female inflorescences), and some of these branches terminate in a male spike with a tassel. He suggested that this lateral branch shortened in a single "catastrophic" event, creating a stem with its male spike and becoming the maize ear. Iltis has since abandoned this hypothesis, largely because it was unsupported by new genetic analyses. But as both he and Doebley make clear, the idea was nonetheless extremely productive because it suggested an important transformation in teosinte: the conversion and condensation of entire lateral branches of teosinte into single ears of maize.[19]

According to John Doebley, it is very likely that changes in the expression of the *tb1* gene during domestication were responsible for teosinte's multi-branched architecture, including its reproductive organs, to transform into the more streamlined version we see in maize. The main function of *tb1* is to produce a protein that represses the growth of **primordia**—the cells in plants that trigger new growth at the tips of organs such as stems, branches, leaves, and flowers. In teosinte, *tb1* is either turned off or suppressed in the primary lateral branches so that they can elongate and tassels can grow at their tips. But in teosinte's secondary lateral branches, *tb1* is turned on, which suppresses branch elongation and tassel formation, allowing ear shoots and leaf husks to develop instead. For maize to evolve, people had to select versions of teosinte where the primary branches also had *tb1* turned on, so that, like the secondary branches, their elongation could also be suppressed—allowing ears to form near the plant's main stem. In a sense, the multiple individual ears we see in teosinte (see figure 3.2) were combined in early maize to form a single ear as a result of *tb1* being turned on (or at least more frequently expressed) in both the primary and secondary lateral branches.

A maize ear is consequently a combination of all the individual ears, tassels, and spikes of the original teosinte architecture. Iltis's insight was that the *teosinte branched1* (*tb1*) locus controlled the branch growth and that by selecting versions of the plant that displayed repressed branch growth, the earliest farmers must have encouraged teosinte's individual separate ears, which are spread out along the lateral branch, to eventually become condensed into a single ear, which is attached to the main stem of the plant with a vestigial foreshortened branch (see figure 1.1).

The single long leaf that, in teosinte, covers the ear stayed associated with the shank, or stem, in maize ears. But, because all of the teosinte ears condensed into a single cob, the protective leaves, themselves condensed into the tightly packed shank nodes, spiraled around the base of the cob, creating a protective husk. One way of thinking of the amazing maize ear with its shank and husk is that it contains all the parts of a teosinte branch—simply with the branch itself shrunken and shortened so that the fruits and leaves that would normally be spread out along the branch now must all grow on top of one another.[20]

Other Genes of Interest

In addition to *tga1* and *tb1* (which are both genes of large effect that control key traits that were important in maize's domestication process) researchers have located several other genes of smaller—and in some cases even unknown—effect that control trait differences among the different varieties of maize. In a recent review Doebley and his colleagues catalog the major genes of large and small effect in a range of domesticated plants, including maize, rice, tomato, wheat, pea, and cauliflower.[21] They include a list (table 8.1) of the major maize genes that have been identified, along with descriptions of their molecular and phenotypic functions. What I find fascinating about this list is that so many of the genes and their functions have only recently been discovered. And new molecular genetic research is narrowing down the question of whether these genes were actually selected for during the domestication process. By the 2006, genetic analyses had shown that about sixty maize genes had been selected for.[22] These include genes that regulate kernel color, starch type, sweetness, inflorescence structure, and a host of other functions, as well as the genes of large effect already mentioned, *tga1* and *tb1*. New work since 2006 has expanded the list manyfold, and it will take many more years to determine the functions of all the genes in maize that must have been impacted by domestication.

Modern molecular methods allow researchers to screen hundreds of maize genes to look for evidence of reduced allelic variation compared to teosinte. The idea is that there should be about as much variation among genes that were not subject to selection during domestication as can be seen in the wild forms of these genes—in other words, they should only show evidence for as much selection as would occur naturally, without human intervention. Genes that were selected for during domestication because of particular valued characteristics should have

Gene(s)	Molecular and phenotypic function	Controls phenotype?	Evidence of selection?	Causative change
Genes identified as controlling domestication traits				
tb1 (teosinte branched1)	Transcriptional regulator (TCP); plant and inflorescence structure	Yes	Yes	Regulatory change
tga1 (teosinte glume architecture1)	Transcriptional regulator (SBP); seed casing	Yes	Yes	Amino acid change
Genes identified as controlling varietal differences				
c1 (colored aleurone1)	Transcriptional regulator (MYB); kernel color	Yes	Yes	Regulatory change
r1 (colored1)	Transcriptional regulator (bHLH); kernel color	Yes	Not tested	Regulatory change
sh2 (shrunken2)	Pyrophosphorylase; supersweet sweet corn	Yes	Not tested	Transposon insertion
su1 (sugary1)	Isoamylase; sweet corn gene	Yes	Yes	Amino acid change
y1 (yellow endosperm1)	Phytoene synthase; carotenoid content	Yes	Yes	Regulatory change
Genes identified by selection screens targeted at candidate genes				
ba1 (barren stalk1)	Transcriptional regulator (bHLH); plant and inflorescence structure	Candidate	Yes	n/a
ra1 (ramosa1)	Transcriptional regulator (MYB); inflorescence structure	Candidate	Yes	n/a
su1 (sugary1), *bt2* (brittle endosperm2), *ae1* (amylose extender1)	Starch biosynthetic enzymes	Candidate	Yes	n/a
Genes identified through untargeted selection screens				
zagl1 (zea agamous-like1)	Transcriptional regulator (MADS)	Unknown	Yes	n/a
ca. 47 genes	Varied functions, including auxin response, growth factor, cell elongation protein, F-box protein, heat shock proteins, hexokinase, kinase, methyl binding protein, steroid biosynthesis, transcription factors, amino acid biosynthesis, and circadian rhythm	Unknown	Yes	n/a

SOURCE: Doebley, Gaut, and Smith 2006.

only the alleles (versions of the genes) that produce the desired characteristics and not the full range present in neutral genes (genes that were not targeted during selection) found in wild teosinte.[23]

In 2005 molecular biologists Stephen Wright and Brandon Gaut described some of the advances in genetic testing technology and the methods that have allowed the kind of screening mentioned above.[24] They estimated that in maize approximately 24 percent of genes that are of agronomic interest were subject to selection. But they cautioned that this might be too large an estimate because of the current bias for looking at genes of interest rather than the whole genome.

One of the outcomes of the focused research that has been conducted so far is that scientists have concentrated on traits that are of agronomic importance in today's economies. We don't know for certain what characteristics of the teosinte plant were of interest to early farmers, so it is difficult to be sure what to look for in the present-day gene screens. Therefore large-scale, whole-genome screening (a topic we will examine in more detail below) is likely to be much more productive in discovering traits that were selected in the past—thousands of years before maize evolved into the plant we now know.

Researchers are finding that some of the genetic loci of interest are linked to neighboring regions on the chromosome, and so it may be that selection in one locus may have indirectly led to change in other loci. This process is called hitchhiking. Is it possible that some of the genetic loci selected inadvertently through hitchhiking eventually became more important than the original genes of interest? One of the ways of determining which genes were of interest in antiquity is to look for genetic variation in ancient maize remains. This is an exciting new direction in archaeogenetics, which is the study of ancient DNA.

DOCUMENTING THE GENETIC CHANGES
IN ARCHAEOLOGICAL SAMPLES

The possibility of discovering ancient DNA first captured the public's imagination with Michael Crichton's novel *Jurassic Park*—later made into a blockbuster movie of the same name.[25] The premise that animals and plants that had been extinct for millions of years could be resurrected using fragmentary nuclear DNA is science fiction par excellence.[26] Even so, the recovery and amplification of ancient nuclear DNA from one of our extinct Neanderthal cousins—a 38,000-year-old individual from the Vindija Cave site in Croatia—has demonstrated that new genetic technology has enormous potential for answering puzzling questions about past biological populations—of all sorts.[27]

Although perhaps not as exciting as reconstructing the extinct Neanderthal genome, the ability to recover ancient DNA from the archaeological remains of our earliest cultivated plants holds enormous scientific promise. In the early 1990s Pierre Goloubinoff, Svante Pääbo, and Allan Wilson recovered gene segments from archaeological samples of maize thought to range in age from several hundred to several thousand years.[28] The ancient samples they examined came from three specimens—two from Peru and one from northern Chile. They chose to look for and extract a gene segment that coded *alcohol dehydrogenase2* (*Adh2*), an enzyme that allows the conversion of some sugars to alcohol—perhaps used by plants to prevent the proliferation of microbes that could be harmful to them at times when sugars are prevalent in their stalks and seeds. The alleles of *Adh2* in the archaeological samples were compared with those extracted from modern samples of maize, teosinte, and *Tripsacum*. More than anything, this study demonstrated that ancient maize DNA could be recovered, and it opened the door to more comprehensive studies.

As mentioned at the beginning of this chapter, a landmark study of ancient maize DNA was published by Vivian Jaenicke-Després and her colleagues in 2003. In the study, the team looked at a sample of maize remains from two archaeological sites: six cobs from Tularosa Cave in New Mexico and five cobs from Romero's Cave and Valenzuela's Cave in the Ocampo region of Tamaulipas, Mexico. Their study was ingenious in several respects. First, they decided to look for ancient DNA representing three genes that, based on the genetic studies of modern maize and teosinte discussed in the previous section, were known to have been targets of selection during domestication: *teosinte branched1* (*tb1*), *prolamin-box binding factor* (*pbf*), and *sugary1* (*su1*). *Pbf* is a gene "involved in the control of storage protein expression in maize kernels," and although its function was not well understood at the time, it was thought to play a role in both the quality and quantity of protein in cereal grains.[29] *su1* is implicated in the production of starch in maize kernels and is thought to regulate the relative proportions of two starch molecules: amylose and amylopectin. The characteristics of these starches greatly affect the properties of the grain and may determine how it can be processed into foods—for example, the quality of starch stickiness influences how maize reacts when it is used to make tortillas.

Because ancient DNA degrades with the passage of time, it is often fragmented, and the long strands have often broken down into smaller chunks. Still, these pieces can reveal diagnostic sequences of base pairs for individual genes and can contain enough information to distinguish among the various alleles of particular genes.[30] Under domestication the allelic diversity of a species is reduced because early agriculturalists select the alleles that yield the characteristics that are of greatest interest to them. By looking at genes known to be the targets of domestication, Jaenicke-Després and her team of researchers could easily determine how allelic diversity in the target genes of the archaeological samples compared with wild teosinte (high allelic diversity) and modern domestic maize (low allelic diversity).

The second ingenious aspect of their study was the use of AMS dating to determine the age of each of the cobs whose DNA was analyzed. This allowed them to identify when and in what order different alleles in the three gene loci became common in the population. The six cobs from Tularosa Cave in the Mogollon Highlands region of New Mexico ranged in age from about 660 to 1870 cal BP. One important discovery was that the oldest cob was found in the same excavation square and level as the youngest cob—demonstrating that the mixing of materials in the cave would not have been detected if each cob had not been individually dated. If the team had assumed the 1,870-year-old cob was the same age as the younger cobs (or vice versa) based solely on the fact that the cobs were found in the same layer, they would not have been able to discern directional changes in the genetic characteristics of the maize.[31] The five cobs from Romero's and Valenzuela's Caves were much older, ranging from about 2350 to 4405 cal BP. The total span of time represented in both regions is almost 4,000 years—giving us a comprehensive view of maize domestication from northeastern Mexico to the southwestern United States.

Jaenicke-Després and her colleagues discovered that these eleven ancient cobs tell us a great deal about the nature and timing of the domestication process. All of the cobs had only one allele for *teosinte branched1*: the allele *Tb1-M1*, which is the version of the gene that suppresses the growth of lateral branches. Teosinte, in comparison, has other versions of the gene—for example, *Tb1-M2*, which occurs in up to 64 percent of most wild populations and allows for multiple branches to grow off the main stalk of the plant. This means that by at least 4,400 years ago the farmers who had moved maize into the Tamaulipas region were growing a subset of the original teosinte population that had lost the ability to branch—demonstrating that domestic maize had passed through a genetic bottleneck, reducing its allelic variability.[32] Remember, the gene *tb1* is also implicated in condensing the secondary branches so that the teosinte ears pile up to form a single cob.

The other major discovery the study made was that all the cobs showed only one of the two alleles for *prolamin-box binding factor:* the version of the gene called *Pbf-M1*. In other words, the variation known to exist in present-day populations of teosinte, which have both alleles (87 percent have the *pbf-M2* variant, while only 13 percent have the *Pbf-M1* variant), had been reduced to just one. This change must have also been in place by 4400 cal BP, and it shows that the earliest maize farmers may have been selecting plants that had desirable storage characteristics for seed protein to seed the following year's crop (although precisely what these characteristics are is not yet well known).

Finally, and perhaps most importantly, Jaenicke-Després and her team found that a gene controlling the way starch forms in the kernels, *sugary1* (*su1*), was not as fixed as the previous two genes. Modern maize has two alleles, *su1-M1* and *su1-M2*, which occur at high frequencies (30 percent and 62 percent, respectively), along with several other alleles that occur in very small frequencies. But these two main alleles are relatively uncommon in wild teosinte (occurring at only about 7 percent). Instead, eight other alleles that are very rare in maize make up about 75 percent of teosinte's *su1* variants. The Romero's and Valenzuela's Caves cobs had only *su1-M2* (the most common *su1* allele in many modern lines of maize), so we know that by 4,400 years ago maize kernels had the starch properties of their modern counterparts. In contrast, the younger Tularosa Cave cobs had primarily *su1-M1*, the less common of the two main *su1* alleles in modern maize. Furthermore, some of the cobs had an allele that is not found in modern maize but is present in teosinte (*su1-T1*). Jaenicke-Després and her colleagues think this might suggest that the genes controlling starch characteristics were not fully fixed by 2,000 years ago. It might also suggest that the early maize that made its way to the southwestern United States diverged from the maize that spread to northeastern Mexico before 4,400 years ago. Early maize probably moved outside of its homeland after it had acquired the desirable traits of *tb1-M1* and *pbf-M1*—a very early bottleneck indeed. Later cultivators in different regions selected variants of the *su1* allele, and they likely selected them at different rates. Farmers in Tamaulipas, Mexico, for example, selected strains of maize that were high in *su1-M2*, while later farmers moving into northern Mexico and the US Southwest selected for maize with higher frequencies of *su1-M1*. We still do not know what, exactly, were the advantages and disadvantages of these different alleles, but it is clear that the earliest maize farmers were very interested in them.

Current and future genetic analyses of archaeological maize samples are going to transform our understanding of the plant's spread and of which characteristics interested ancient farmers. As maize spread out of its initial home range, it had to adapt both to its human cultivators and their cultural preferences and to its new environments. It may be that the diffusion of early versions of maize into the northern regions of Mexico and the southern United States had to await the gradual selection of traits that would permit it to survive a range of new conditions: colder temperatures, more daylight hours during the summers, decreased rainfall, and different pests—both insects and microbes. By isolating in archaeological samples other genes that were known targets of selection—as well as genes that weren't—it will surely be possible to generate new hypotheses explaining the timing and direction of genetic changes during the domestication process. Several researchers have pointed out, however, that we do not have many existing collections on which to perform these analyses.[33] While this may be true for the earliest known samples, there are many younger, well-preserved remains being found in South America, especially in the arid regions of coastal Peru and northern Chile, that promise to yield new genetic information. Future studies of these plant remains will provide some of the most interesting glimpses of the timing and evolution of maize under domestication.

ESTIMATING THE AGE AND RATES OF MAIZE'S GENETIC EVOLUTION

What can genetic evidence tell us about the timing of teosinte's domestication and the evolutionary changes that took place in maize? We have just seen that the genetics of present-day plants and the reconstructed genetics of ancient archaeological samples give us many new insights into maize's history. Geneticists are also able to use their biological data to peek at maize's molecular clock and estimate the timing of the plant's evolution. The methods are based on complex calculations of rates of mutation in the DNA—and these rates are based on assumptions that are difficult to confirm because they cannot be directly observed over long time spans. But the strength of this approach is that it allows geneticists to provide a best guess for the elapsed time since teosinte separated from its descendants that were transformed by domestication—the descendants that eventually became modern maize. The estimate can then be compared with other estimates that are based on methods such as radiocarbon dating.

One way that geneticists have estimated maize's rate of mutation is as follows.[34] To calculate the number of elapsed generations since the two subspecies fully split into isolated populations, geneticists needed to estimate the mutational rate of a sample of genes. They selected thirty-three genetic loci for comparison. The question is, however, How does one determine the rate of mutation among these thirty-three loci (or any others for that matter)? Yoshihiro Matsuoka and his colleagues decided to grow several generations of maize and trace the number of mutations that occurred in their thirty-three target loci over this span of time. This was done by growing eighty-six plants for eleven generations (one generation is assumed to represent one year) and then comparing the genotypes at the thirty-three loci between offspring and parents for each generation. The plants were "selfed" (self-pollinated), so that rates of mutation could be observed across the eleven generations of each of the eighty-six original plants. They discovered that thirteen novel alleles appeared during this time span, and they used this observed mutation rate to extrapolate an overall mutation rate.

The estimated mutation rate allowed the team to calculate that *parviglumis* and Mexican maize diverged 9,188 years ago, give or take about 3,500 years (i.e., with 95 percent confidence, in statistical terms). This 7,000-year "window" of divergence is not particularly helpful, since we already know that the earliest macrobotanical remains of maize from Guilá Naquitz in Oaxaca date to approximately 6,250 years ago.[35] Although these early cobs fall within the span suggested by the molecular estimate, they narrow the window to sometime between 6,250 and 9,000 years ago. So although the molecular estimate doesn't contradict the archaeological evidence too much, it isn't really precise enough to narrow down the time frame in the absence of new archaeological data.

The previously mentioned study of teosinte glume architecture (*tga1*) by Wang and his colleagues takes a slightly different approach to calculating the divergence of maize and teosinte.[36] Using a recently developed technique, they estimated the time in generations (each generation is considered to be one year) since the fixation in the population of the favored allele—in this case, *Tga1-maize*.[37] With an assumed population size of one hundred thousand plants, they calculated that *tga1* was fixed (that is, dominant for *Tga1-maize*) in the population by about 10,000 years ago. This is somewhat earlier than the previous estimate of about 9,188 years ago, but, taken together, these calculations suggest that the earliest maize plants, the ancestors of all those that followed, may have been around for 3,000 years or more before they appeared at Guilá Naquitz Cave.

MAIZE HAPMAP

Ten years ago it seemed fairly clear that the technology of genetic sequencing and analysis was about to explode. And that is exactly what has happened. The cost of sequencing genomes is falling while the speed, accuracy, and completeness of coverage steadily rises. It is now possible to analyze hundreds of thousands of SNPs (single-nucleotide polymorphisms) from a single individual—plant, animal, or microbe— and to do so in a fraction of the time and cost that it used to take. Some of the key scientists responsible for many of the previous maize genetic studies we've looked at have now teamed up to tackle the problem of identifying and decoding the entire *Zea mays* genome—including domesticated maize and its wild relatives.[38] Under the banner of the Panzea Project, the researchers have compiled an astonishing amount of genetic information about the members of the *Zea* genus and have made it available to the world on publicly accessible databases.[39] They have also published a series of papers describing their detailed analyses of vast numbers of SNPs—one outcome of which has been to allow increasingly accurate assessments of maize's family history.

The first landmark study, published in 2009, unveiled the HapMap project—called HapMap Version 1—which allowed Edward Buckler and fellow maize geneticists from a half dozen institutions to describe the maize **haplotype** in the greatest detail so far. Haplotypes characterize the genetic variation in the chromosomes of any organism, based on the similarities and differences in their SNPs when compared with other members of the same or related species. Similar studies are being conducted for many other species, including humans, and all of them are aimed at better understanding the full scope of genetic variation within populations of organisms. In the maize HapMap study, the research team was able to record 3.3 million SNPs and **indels** (small insertion/ deletions) on twenty-seven inbred lines of maize.[40] One of their key findings was that there was a great deal of evidence for multiple and large-scale selective sweeps (places where genetic diversity had been reduced in various regions of the individual maize samples tested), suggesting that this could be related to maize's adaptation to geographic variation as it moved into different environments throughout the Americas.

Three years later, in 2012, HapMap Version 2 was released, and it was summarized in two papers published in *Nature Genetics*.[41] The HapMapV2 team looked at 103 samples of maize and teosinte lines, along with one *Tripsacum* specimen as an out-group, enabling the characterization of the genotypes of 55 million SNPs and indels. The enormous amount of data generated by these studies is available online and can now be used as a reference in future research. For example, maize geneticists can now compare the SNPs recovered from fragmentary archaeological specimens to this reference database, making the rapid assessment of SNP variability for thousands of genes in ancient maize samples possible—something that could previously only be done with a handful of target genes, as we saw in the 2003 study by Jaenicke-Després and her colleagues. Still, it will be a long time before the functions of these newly discovered genes are determined and before it is established whether or not they were actively selected for during the long process of domestication and regional adaptation.

Taken together, the new research on the genetics of maize evolution has solved a number of puzzles and simultaneously led to several important new questions. Thanks to the innovative work of the botanists and geneticists just discussed, we now have a much clearer understanding of

(1) where maize was originally domesticated—in the Balsas River region;

(2) what maize's progenitor was—*Zea mays* ssp. *parviglumis*, with some introgression from ssp. *mexicana;*

(3) when the first significant genetic selection took place as a result of human intervention—about 9,000 years ago, give or take several centuries;

(4) what kinds of changes were most significant in the early domestications process—glume architecture, branching, starch and protein quality, and several others; and

(5) what the rough timing and directionality of the spread of maize throughout the Americas was—both northward, into northern Mexico and the southern parts of the United States, and southward, into southern Mexico, Central America, and eventually South America.

While these answers take us much closer to understanding the genetic origins and spread of one of the modern world's most important plant crops, there are more questions ahead—questions that link us to the social, cultural, and technological aspects of maize's history. To understand the cultural impacts of this key crop and its changing roles through time, we need to dig back into the archaeological record and examine a different type of evidence—the tools and other objects that people used to store, cook, process, and serve their precious crop. In the final chapter of this volume we will explore questions such as: What can the archaeological record tell us about how people used maize? What was its nutritional and symbolic importance? How did different peoples in various regions of the Americas mold the plant to their needs and how were they, in turn, affected by its enormous power?

CHAPTER 9

Daily Tools and Sacred Symbols

When excavating the Postclassic Maya site of Canajasté in the early 1980s, my crew and I found many examples of stone tools and pottery vessels used for grinding and cooking maize. These tools were easily recognizable as such because, although they were made and used during the thirteenth and fourteenth centuries, they are almost identical to the maize cooking utensils that the local population still uses today—and also those of many peoples living throughout rural Mexico, Central America, and beyond. Some ceramic pots, called *ollas,* were used to boil maize kernels, beans, and many other foods; some were used as colanders for washing the lime-soaked maize after boiling; and others were used as griddles to cook tortillas. This basic cooking toolkit could be found in the kitchens of most ancient Mesoamerican households by the beginning of the Classic period, some 1,700 years ago, and even earlier, during the Preclassic period, at some ancient settlements across the region.

But this toolkit, designed especially for preparing and cooking maize, was not present during the Early Preclassic—at least not in all regions. On the Pacific coast of Chiapas, for example, where my colleagues and I have been excavating household deposits dating to between 3,900 and 3,000 years old, we find a great deal of pottery, but very little of it would have been suitable for boiling foods, including maize, over open fires. We don't find griddles for cooking foods either, and rarely do we encounter the types of grinding stones that are so commonly associated with maize preparation in later periods. This doesn't mean maize wasn't being used—in fact it was often the most common plant species that we found in our search for ancient botanical remains at the Early Preclassic sites we excavated in the Mazatán region of Chiapas, Mexico. It does mean, however, that maize was not being processed in the ways it was during subsequent millennia. Was this because maize wasn't yet very productive? Or did it simply have low priority among the inhabitants of the region? Or were they eating it in ways that didn't require long-term storage or prolonged cooking? We saw in chapter 7 that maize was present in the diet of the first villagers in the region, as detected by isotopic analyses, but that it did not appear to be the dominant food source. What are the implications of this pattern of an early adoption of maize but a slow transition to its being used as a dietary staple, and what does this have to do with the way maize was processed and consumed? Perhaps the examples of cooking technologies from the two ends of the spectrum—the Early Preclassic peoples on the coast of Chiapas and the Late Postclassic peoples in the interior of the state—can help give us an important insight into the transformation of maize from an occasional food source to a dietary mainstay.

The appearance of a developed maize-cooking assemblage of pottery vessels and grinding tools in residential deposits at archaeological sites by about 3,000 years ago points to the increasing importance of maize in these economies, and it tells us that by this time maize consumption was becoming a significant part of most peoples' diets—in other words, it was becoming a food staple and not simply a supplement. Befitting its dietary importance, we might also expect to see graphic and symbolic representations of maize in the archaeological record. Indeed, depictions of maize and maize deities become increasingly common throughout Mesoamerica beginning during the Preclassic period, about 3,000 years ago. In this chapter we will look at the archaeological evidence for maize cooking technology in the Americas and see when and where this technology is accompanied by changes in the symbolic realm. When do people start depicting the sacred nature of maize and its connections to life, cosmology, and the gods?

CONSUMING MAIZE

There are many ways of eating maize, but with the exception of ingesting immature green ears as one might enjoy any raw vegetable, almost all of them involve some sort of cooking and processing. Whether working with freshly picked ripe ears or dried kernels that have been kept in storage for many months, maize must be cooked to make it palatable and to release its nutrients. Fresh ears of corn on the cob can be steamed in pits, roasted over an open fire, or simmered for a few minutes in a pot of boiling water—always yielding delicious results. Dried maize is much more difficult to process, unless it is popcorn, in which case it can simply be heated by an open fire until the kernels explode and the tasty morsels can be gobbled up. But otherwise, there are only two main ways of cooking dried maize. Kernels can be either dry ground into coarse flour and then boiled, or boiled whole and then wet ground. A variant of the first option is to toast the kernels prior to grinding them— a technique often used to produce a ceremonial drink made from dried maize. Either way, in the preindustrial era this required two types of technology: grinding stones and boiling pots.

When dry grinding maize before cooking, the hard, desiccated kernels would be ground on a trough- or mortar-shaped stone slab so that the meal stayed in place, making it possible to continuously grind the particles to an ever smaller size without having them fall off the edge of the grinding surface. The dry ground meal could then be added to water and boiled into a porridge-like mass. In contrast, when the whole grains are cooked before grinding, the boiled and softened kernels could be securely placed on a flat stone and ground into a soft dough (or *masa* in Spanish) that would remain in place as it was worked back and forth across the grinding surface.[1]

GRINDING STONES

Throughout Mexico and Central America cooked maize was ground using two milling stones: the mano and the metate. The mano, held with either one or two hands, was used to grind the kernels against the metate's large, flat surface. These stones were, and in many communities still are, essential tools in every household's kitchen—a basic requirement for making *masa*, which can then be used in a number of foods, such as tortillas, tamales, and a porridge-like gruel (figure 9.1). Farther to the north, in northern Mexico and the southwestern United States, manos were often smaller and shaped to fit in trough- or basin-shaped metates, on which the dry kernels were ground into a coarse flour or meal. After boiling the meal, the end result was similar to *masa*—and the cooked maize could be prepared as cakes or tamales, or added to stews with dozens of other ingredients. In other parts of North America people would grind the maize using large hardwood mortars

FIGURE 9.1. Apart from the hearth, the maize preparation counter is the most important part of every Tzeltal Maya kitchen in the town of Chanal in Chiapas, Mexico. The large stone metate sits on the table to the left, and atop it is a gourd tortilla container. To the right, clamped to the table, is a metal maize hand-crank grinder (wrapped in a cloth). Under the table, propped against the kitchen wall beneath the metate, sits a large ceramic *comal*, or griddle, used for cooking tortillas and toasting other foods. To the right, on the kitchen floor, is a large earthenware *olla* to boil maize and beans. A large thermal flake has spalled off the pot near its shoulder, where it has come in contact with hearth flames during many hours of cooking over an open fire. In the mid-1970s, when this photo was taken, virtually every house in Chanal had this assemblage of maize cooking and grinding utensils in their kitchens. (Photograph by Michael Blake)

and pestles—to the same effect as grinding stones. Since wood doesn't preserve as well as stone in most archaeological contexts it is harder to trace the antiquity of such wooden grinding tools.

Stone manos and metates and stone mortars and pestles preserve exceedingly well in the archaeological record and are commonly found in ancient settlements around the world, spanning at least the past 10,000 years. In the Americas these specialized grinding tools, in one form or another, date back to before the early Archaic period, when people ground many different types of nutritious seeds in order to more easily consume them, including—in Mexico at least—ancestral teosinte.[2] However, the specialized manos and metates that are most closely associated with maize grinding practices today only appeared in the archaeological record by about three to four thousand years ago in most parts of Mesoamerica and much later in other regions to the north and south.[3]

In a recent study of the stone grinding tools recovered from excavations at the Maya site of Santa Rita Corozal, located on the Caribbean coast of Belize, archaeologists have documented the tools' distribution in household deposits spanning the period from the Early Preclassic to the Late Postclassic.[4] At this site many different types of grinding stones were present, including flat-surface, trough, basin, and concave metates, along with single- and two-handed cylindrical manos. This pattern is quite typical of grinding stone technology in that there is a good deal of variation in the type of grinding stone styles used—but even so there is a gradual shift over time to a higher frequency of flat-surface metates and two-handed manos. This appears to be strongly correlated with the shift to an increasing reliance on maize as the primary dietary staple throughout Mesoamerica during the Classic and Postclassic periods. At sites such as Santa Rita Corozal the mix of metate and mano shapes indicates that they were used for a variety of grinding tasks beyond straightforward maize processing—perhaps for grinding other foods or even non-food materials such as minerals used to make pigments.

At the Postclassic Maya site of Canajasté I recovered 135 fragments of manos and metates, mostly discarded pieces of flat-surface metates and two-handed manos—no complete ones turned up in our excavations.[5] For the most part the fragments showed signs of wear and tear, meaning that they had broken during use or perhaps by accident. Most of the fragments had been incorporated into refuse deposits or the construction fill used to build houses. The key pattern in the late period deposits at sites like Santa Rita Corozal and Canajasté is that even though manos and metates were relatively rare in each household— probably only one or two sets per family—they would eventually break and be discarded in the household trash. Over the course of several generations these fragments would accumulate in the archaeological record of the village households (figures 9.2 and 9.3).

This contrasts dramatically with the patterns we see in the Early Formative period household deposits on the Pacific coast of Chiapas. At sites such as Paso de la Amada, in the Mazatán region, we find many small grinding tools but few that could be interpreted to have been manos and metates in the sense that we see in later time periods. Analyses of the artifact assemblages from many Late Archaic and Early Formative sites in Chiapas and elsewhere in Mesoamerica show that such tools dedicated to the grinding of maize were rare.[6] In fact my colleague

FIGURE 9.2. Metate fragments from house floors and middens at Canajasté, a Postclassic period Maya center in the Upper Tributaries region of the Grijalva River Basin in eastern Chiapas, Mexico. One fragment (top center) is the foot of a metate that must have been very much like the one shown in Figure 9.1. Another fragment (bottom center) is from a slab metate, which became highly polished from years of use before it was broken and discarded. (Photograph by Michael Blake)

FIGURE 9.3. Mano fragments from Postclassic period house floors and middens at Canajasté. After they were broken, people continued to use such fragments as hammers. (Photograph by Michael Blake)

John Clark, in his summary of the early grinding stone technology recovered at our excavations at Paso de la Amada, states that they were "(1) present in limited numbers, (2) designed for small grinding tasks and probably a range of tasks, and (3) generally little-worn. . . . The Early Formative tools appear unspecialized, inefficient, and light weight."[7] They increase in frequency and diversity by the end of the Early Formative and become increasingly common at the beginning of the Middle Formative, around 3,000 years ago.

In many ways, this Early Formative grinding toolkit is not so different from that seen in earlier times throughout Mesoamerica. The preceding Archaic period milling stone assemblage is well represented at sites such as those excavated by Richard MacNeish in the Tehuacán Valley and by Kent Flannery in the Oaxaca Valley. At sites in these regions, both of which are famous for yielding some of the earliest known examples of maize in the Americas, the grinding stones include shapes and sizes that could have been used for processing many different types of seeds and other materials.

COOKING POTS

Ceramic technology is often closely linked to food processing in settled villages for two reasons: first because pottery can easily contain liquids and allow them to be heated over fires, and second because they are durable and can withstand a good deal of movement. However, among nomadic peoples around the world, pottery is rarely used for storage, transport, and cooking because pots are much more easily broken and much heavier than baskets, bags, and wooden containers. It is often only when people begin to live in permanent settlements that they begin to rely on ceramic technologies. There are, of course, exceptions to this pattern—for example, in cases where sedentary villagers do not adopt or develop ceramic technology, or when nomadic or seminomadic people do. But as in many other parts of the world, in the Americas there is a close correlation between increasing sedentism, increasing reliance on agriculture, and increasing use of ceramic technology.

When it comes to using maize, ceramics provide many important technological functions: storage, boiling, and serving. While many other types of containers may work equally well for storage and serving, ceramics are the best non-metal technology for cooking dried, rather than fresh, maize kernels. Cooking pots could be used to boil the maize in water for long periods, usually over open fires. Examples of boiling

FIGURE 9.4. A scene from Fray Bernardino de Sahagún's *General History of the Things of New Spain*—commonly known as the Florentine Codex—showing an Aztec woman singing or saying a prayer to the maize she is pouring into a cooking pot (book I, folio 347 right).

pots from different regions of the Americas usually show some of the same characteristics. They are generally large enough to hold a sufficient quantity of grain to feed a household for a day, and they can be tempered with materials such as fiber, shell, or calcite to help prevent thermal shock during prolonged exposure over open fires.[8] They are usually shaped so as to allow maximum exposure to heat, with incurved necks so they can contain boiling liquid. Throughout Latin America these cooking vessels are known as *ollas* (figure 9.4).

When preparing dried maize kernels for cooking, the grain is usually soaked in water overnight. Next an *olla* containing maize and water is set near an open fire for a few hours until the kernels are thoroughly cooked. The water is then drained off and the cooked maize is rinsed and ground using milling stones to form a dough that can be prepared for consumption in many different ways. Before discussing the methods for turning cooked maize into meals, we need to look at an important aspect of the chemistry of maize preparation, because this has an enormous impact on nutrition and health.

NIXTAMALIZATION

Nixtamalization is the process of soaking or boiling maize kernels in an alkali solution in order to improve the availability of nutrients, especially proteins, for human consumption. "Nixtamal" comes directly from *nextamalli,* a word in Nahuatl—the language of the Aztecs, which is still widely spoken in central Mexico today—that combines the words for lime (*nixtli*) and maize dough (*tamalil*).[9] Most readers will already be familiar with the word *tamalil,* since it is the root of the name of a Mexican culinary delight—the tamale—made, of course, from maize dough, wrapped in leaves and steamed.

In the early 1970s the biological anthropologist Solomon H. Katz and his team published an influential paper exploring the implications of nixtamalization and its ability to overcome the nutritional problems caused by the fact that maize "is deficient in the essential amino acids lysine and tryptophan, and in niacin, a member of the vitamin B complex."[10] Building on the earlier work of nutritionists and chemists, the team described how boiling maize kernels in a solution containing lime or other sources of alkali helps to release, through a chemical reaction, much more of these essential amino acids—up to 2.8 times more—than would be available by simply boiling maize in plain water or cooking by other means.[11] Even though the total proportion of protein to weight is decreased by this process, available lycine and tryptophan are increased—and these are key to a healthy diet.

This cooking process is particularly important when maize comprises a significant proportion of the overall diet. People who rely on maize that has not undergone the nixtamalization process are at risk of protein and vitamin B deficiency—which can lead to growth problems and diseases such as pellagra.[12] Pellagra and other maladies caused by an overreliance on maize have caused serious health problems in populations around the world. But once the connection between maize-heavy diets and vitamin B deficiency was worked out early in the twentieth century, it has been relatively easy to avoid these diseases. We know that ancient Americans must have solved this problem—whether intentionally or by accident—because otherwise they would not have been able to rely on maize as one of cornerstones of their diet. Just when they did this can be inferred in part by the appearance in the archaeological record of technologies and techniques to nixtamalize maize so that its essential nutrients could be released.

The nixtamalization process is key in understanding the history of maize's spread because it provides a clue as to when maize made the shift from being just one food source among many to becoming a dietary staple. The work of Katz and others shows that the increasing significance of maize in people's diets wasn't simply the result of producing larger and more productive ears of maize; equally important was the cultural knowledge of how to prepare it—the development of cooking technologies that could unleash maize's nutritive capacity.

During some of the initial stages of maize use, people may not have needed to worry so much about its potential nutritional deficiencies because it was only a minor part of the overall diet. The mix of other foods in people's varied and abundant diets could easily have supplied the essential amino acids that people needed. However, when maize took on an increasingly significant role in providing calories for hungry populations, people were exposed to serious nutritional deficits as a result of the reduction in lycine and niacin in their diets, and this could lead to dangerous illnesses.

As part of their study Katz and his team surveyed early descriptions of maize-processing techniques in previously published ethnographies from the Americas to determine how widespread nixtamalization was prior to the arrival of Europeans (map 9.1). In a sample of about fifty cultures all over the Americas they found that maize nixtamalization was, with one or two minor exceptions, used only in continental North America.[13] It was almost completely absent among the societies they sampled in lower Central America and South America. They make the point that throughout Mesoamerica and North America, where nixtamalization was practiced in one form or another, peoples relied on maize as a subsistence staple. This suggests a radically different history of maize use between the two regions. In the north maize may have spread from its original homeland in western Mexico because it was seen primarily as a food source. In contrast, as maize moved southward into South America, it may have been adopted for different reasons. One possibility is that maize in South America, perhaps from its earliest times, was used primarily as either a supplementary food or as a carbohydrate source for making *chicha*—maize beer.

COLANDERS AND GRIDDLES

Returning to the examples of cooking utensils recovered from my household excavations in Chiapas—from Early Preclassic sites on the Pacific coast and Postclassic sites farther inland—we can see major changes in cooking technology. At Postclassic Canajasté there were many different types of ceramics used in maize processing. In addition to *ollas* for cooking maize, there were two other common types of pottery utensils: colanders and griddles. Colanders, locally called *pichanchas,* are essentially

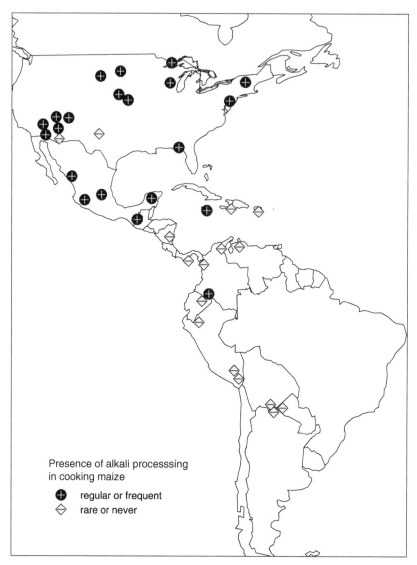

MAP 9.1. The distribution of nixtamalization, based on a survey of ethnographic and historic descriptions of alkali processing used in maize cooking compiled by Solomon Katz and his colleagues in 1974. (By Michael Blake and Nick Waber)

FIGURE 9.5. Steam rises from boiled maize that has just been poured into a colander (*pichancha*) used for rinsing the limewater off the cooked kernels prior to grinding. This *pichancha* is set on a three-pronged wooden branch sunk into the patio outside a household kitchen in the Tzeltal town of Chanal in the highlands of Chiapas. Note the accumulated hearth ash and sludge from countless episodes of cooking maize. (Photograph by Michael Blake)

ceramic strainers. They are a special type of *olla* that is made with regular, closely spaced perforations over most of the body of the pot, so that when boiled maize kernels are poured out of the cooking pot and into the colander all of the cooking water can be rinsed out. This brings us back to nixtamalization. As part of the nixtamalization process the cooked maize must have the alkaline cooking water rinsed out of it. When doing this on a daily basis it makes a great deal of sense to invest in making a sturdy strainer—and pottery works wonderfully well in this regard. In Chiapas today there are two main types of ceramic *pichanchas:* the most common is the large-holed variety used for the daily washing of boiled maize, and the rarer variety is a smaller-holed type that is used for straining finely ground and flavored maize gruel called *atole*—used during ceremonial occasions (figure 9.5). It is the larger-holed *pichancha* type that concerns us here, as they are the ones most commonly found in Late Classic and Postclassic time periods, which is when we think processing maize using the nixtamalization process had become extremely important.

Another very common type of pottery found at Canajasté is a circular griddle called a *comal*. *Comales* are used to cook tortillas over an open fire. Like *pichanchas, comales* began to appear commonly throughout

Mesoamerica during the Classic period and increased in frequency during the Postclassic. This combination of ceramic tools—*ollas* for boiling maize in alkaline water, *pichanchas* for straining the cooked maize, and *comales* for cooking tortillas—in combination with stone manos and metates for grinding the cooked maize to make *masa,* provides a technological signature denoting both the importance of maize as a food staple and the importance of the nixtamalization process for ensuring that this staple was nutritious and healthy.

Comales are generally flat ceramic discs with a slightly thickened and upturned rim. Their surfaces are generally smooth, while the undersides can be rough and scraped. They range in diameter from 30 centimeters to almost a meter and are usually about a centimeter in thickness. Often found broken in household trash, like boiling pots, they have a short use-life because they are constantly employed in busy kitchens and being moved back and forth over open fires. In Postclassic period sites like Canajasté there are often dozens of *comal* fragments found in the household trash deposits off to the side of the kitchen or incorporated into the construction fill of buildings. I found two infant burials beneath the floors of two separate houses, and in each a *comal* was used as a lid to cover an *olla* containing the tiny body of the newborn child (figure 9.6). This gives us some sense of the symbolic importance of these vessels, which were used to prepare life-sustaining food and also, occasionally, to protect the spirits of the dead in the afterlife.

Similar ceramic griddles are also found in other parts of the Americas. They are well known in Amazonian South America and the Caribbean, where people cooked thin tortilla-like cakes using flour made from the manioc root (cassava). There is very little ethnographic evidence that the griddles were used to make maize tortillas in these regions—so when they are recovered in archaeological sites they very likely signify the cooking of manioc cakes rather than maize.

In the southwestern United States and northern Mexico griddles were made of both ceramic and stone. Margaret Beck's summary of maize use in these regions shows not only that these griddles, or *comales,* were commonly reported in historic accounts from the region, but also that their ancient counterparts are represented in many archaeological sites throughout the area. She summarized the archaeological evidence for the use of ceramic griddles in the Hohokam culture of Arizona, finding that the Classic period Hohokam didn't start using them until about 900 years ago, when they appear regularly at eleven of the nineteen sites studied.[14] One fascinating aspect of this finding is that it

10 cm

FIGURE 9.6. At the Postclassic Maya site of Canajasté, several infant burials were found beneath house floors. The infants, who may have died in childbirth or shortly after, were sometimes buried in *ollas,* and in one case the small burial pot was carefully covered with a broken *comal.* (Photograph by Michael Blake)

corresponds roughly to the period when *comales* become common throughout Mesoamerica—among the Aztecs, Maya, Zapotecs, and many other peoples in the region.[15]

Even though specialized ceramics for cooking and preparing maize, including *ollas, comales,* and colanders, became increasingly common during the past millennium or so and there are traces of this technology prior to 3000 BP, there is little evidence that it played a significant role before this time. For example, returning to the Early Formative Pacific coast sites, we find that some of the earliest people to live in settled villages on the coast of Mexico both consumed maize and made excellent pottery, yet they did not seem to use any of the specialized maize-processing toolkit. In fact, at all of the Early Preclassic period village sites that my colleagues and I investigated in the Mazatán region, we found very little evidence that people used ceramic vessels for cooking maize—or anything else for that matter. Most of the pottery that people used consisted of beautifully made and often highly decorated serving and storage vessels. Analyses of the Early Preclassic ceramics found on the coast of Chiapas show that most types of pottery may have been initially used as part of displays during the serving and eating of food.[16] Although maize was definitely grown and eaten at sites in the Soconusco region, it does not appear to have been processed using nixtamalization, nor even to have been boiled in ceramic pots, as it was in later periods. Instead we found that foods may have more frequently been cooked in earth ovens or by using heated stones placed in watertight containers that were perhaps made of wood or basketry.[17] We did not recover many mano- and metate-type grinding stones from our excavations either. The grinding stones we did find were generally small mortars and pestles or ones that could be used for milling seeds or pulverizing minerals for pigments. Therefore, on the coast of Chiapas during the time of the first settled villages, when pottery had become common and maize was a regular food item, I think it was highly unlikely that people processed maize using nixtamalization. This suggests very strongly that maize was not yet the important food staple that it later became in that region. This doesn't mean that it was not important in other ways, such as nutritionally and symbolically, but rather that people probably did not derive the majority of their caloric intake from maize. If they had, they most likely would have suffered from severe nutritional problems, such as pellagra. In fact it may have been these diseases that led to nixtamalization becoming increasingly common during the Classic period in Mesoamerica and in adjacent regions of northern Mexico and the United States as cities and towns grew in size and more and more people had to rely on maize as their main source of carbohydrates.

One possible exception to this pattern has been recently discussed by David Cheetham, who has shown that some of the earliest known ceramic colanders—dating from about 3,000 to 2,800 years ago, a period known as the Cunil Horizon—have been found in the Belize River valley in Belize. They disappeared after this period and were slow to re-emerge in ceramic assemblages later in Belize and elsewhere in Mesoamerica.[18]

When do we see the more widespread use of colanders in the archaeological record of Mesoamerica? Although they do occur in Classic period sites, it is really not until early in the Postclassic period, about 1,000 years ago, that the combination of ceramic colanders and flat ceramic griddles show up with regularity.[19] This suggests that the nixtamalization process was necessary for tortilla making, because it gave the *masa* the right consistency to hold together and made it easier to work with when preparing the tortillas. One of the benefits of nixtamalization is that it helps loosen the pericarp, or the tough outer shell of the kernel, so that it can come off during rinsing and grinding, allowing for a smoother consistency when made into dough. Dough made from well-ground maize could be made into balls, flattened, and cooked on a griddle, producing a very different type of food than could made using more coarsely ground *masa*. The careful washing, rinsing, and straining of the cooked maize kernels is what makes it possible to grind them into a dough with just the right consistency to stick together properly when pressed into the shape of tortillas.

In the southwestern United States and northern Mexico people also boiled maize in limewater solution. Whether the maize was boiled before or after grinding, it was a crucial step because the dried maize kernels had been stored over the winter and boiling is the easiest way to render the rock-hard grain edible once again. In her survey of historically documented maize cooking practices among dozens of groups in this region, Margaret Beck lists the many ways people prepared dried maize for consumption. The Hopi people in Arizona boiled or soaked ground maize in juniper ashes to make hominy. Others, including the Tarahumara of Chihuahua and Sonora in Mexico, used the ashes of oak bark or maize cobs, soaking the kernels either before or during the boiling stage.[20] In addition to consuming this cooked maize as hominy, many groups in the US Southwest made it into dough, which they fashioned into tortillas that could be cooked on stone or ceramic griddles. It was also common for people to prepare maize in stews or as balls of dough that were wrapped in leaves and baked in a fire.

There are similar documented accounts of boiling maize in limewater solutions throughout North America—wherever hominy was made and consumed. In northeastern North America, "from southern Ontario to the Gulf Coast and from the Atlantic Ocean to the Missouri River," hominy was the primary means of preparing maize by at least 500 years ago and possibly as early as 1,000 years ago in some regions.[21] Even though maize had been introduced into the northeastern United States by at least 2,000 years ago, it appears to have played a rather minor role in the diet until much later. Available evidence points to its gradual spread northward and eastward, and it eventually made its way as far north as southern Ontario and Manitoba to the west and Rhode Island to coastal Maine the east.[22] Its increasing dietary importance in the northern latitudes was very likely correlated with improved varieties of maize that were developed as the plant gradually adapted to the temperate—and somewhat harsher—environmental conditions that it encountered as it spread as much as 3,500 kilometers north of its tropical homeland.

Myers's summary of the methods many of the peoples inhabiting this enormous region used to process maize into hominy shows a great similarity in technique and technology. One of the important contrasts between the northeastern and the southwestern United States is that in the northeast there was a switch from the use of stone slab metates to large wooden mortars and pestles. The stone metates and manos are thought to have been used to process seeds and nuts from a wide range of native food plants, some of which were domesticated, whereas the wooden mortars and pestles widely described in the ethnohistoric literature were developed to process dried maize into hominy.[23]

Hominy was commonly made by placing the dry maize kernels in a mortar made out of a partially hollowed out log and then pounding the grain and a mixture of wood ash together with a long wooden pestle. The resulting maize meal was sifted to remove the chaff and hull fragments and then boiled in ceramic pots, at which point a range of other ingredients could be added, depending on the recipe. Myers makes the point that this wooden grinding technology is difficult to find in the archaeological record because wood, like most uncharred organic materials, seldom preserves for long in the damp environment of the region. However, the hypothesis does fit with the observations that stone slab metates become less common as maize use increased in importance throughout the region. He also notes that peoples living farther to the west in the plains also made hominy, but they did so by boiling the whole kernels in a lye solution made from wood ash. Unlike their neighbors to the east, who ground the kernels before boiling, or their neighbors to the south, who generally ground the kernels after boiling, groups in central North America, such as the Hidatsa people, would simply rub off the boiled pericarps and then add the whole kernels to stews for further cooking.

All of the maize preparation techniques that include nixtamalization became increasingly important as maize transformed from a supplemental food to a dietary staple. Every one of these methods requires some degree of soaking and/or boiling maize in an alkali solution and, unless they involve the use of a durable technology, such as pottery vessels made from fired clay or grinding utensils made from stone, traces of these processes are unlikely to preserve in the archaeological record. This could make it very difficult to find direct archaeological evidence of nixtamalization. But when we have determined, by other means, that maize was consumed as a primary food staple, we can infer that people must have used nixtamalization, because otherwise they would have faced the risk of malnutrition and the maladies that accompany vitamin deficiencies.

REPRESENTING MAIZE

What causes us to link what we eat and drink to supernatural powers? For the ancient Greeks, the food and drink of the gods were special and sacred. The words "ambrosia" and "nectar" are derived from Greek and are still used in most Western languages to describe perfect foods and drinks—those fit for deities and royalty.[24] When and why do particular foods and beverages become deified? This is a difficult question to answer because in order to trace the emergence of sacred foods in antiquity, we must be able to see signs of their "sacredness" and their uses in rituals and offerings. It may, in some cases, be possible to see traces of ritual offerings in the archaeological record, but these can be difficult to distinguish from day-to-day food preparation and consumption. But ancient representations of the sacred uses of foods and drinks can be very useful, because there is no reason these uses would be represented symbolically, apart from the existence of the ritual practice of consumption or offering itself. If a food or drink is represented artistically and the artistic representation is linked to the imagery of supernatural beings or forces, this can give us a window on the ancient roles of these foods and drinks—ambrosia and nectar—as the Greeks would say. From an archaeological point of view, we can generally discern symbolic practice only when it is represented materially. Of course there are many intangible ways of representing ideas—such as dance, stories, and song—but without direct observations of these practices recorded in oral or written history, we may never have any knowledge of them. However, when abstract, symbolic concepts are represented in material form, even in very ancient times, we have the ability to see such practices, although we may perhaps never fully understand them.

The earliest such clear-cut material representations of the sacred and symbolic importance of maize comes from the Olmec people of southern Mesoamerica, who carved discernable images of maize on stone, wood, and pottery. Some of their most ancient representations are rather simple depictions of ears of maize, while others are much more abstract and powerful images of the Olmec Maize God. Michael Coe linked the representations of maize identified on Olmec pottery and engraved on jade celts with the cleft on the head of many images of a supernatural Olmec entity, the were-jaguar. Building on the earlier interpretations of Miguel

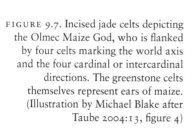
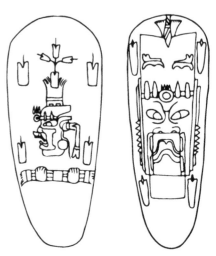

FIGURE 9.7. Incised jade celts depicting the Olmec Maize God, who is flanked by four celts marking the world axis and the four cardinal or intercardinal directions. The greenstone celts themselves represent ears of maize. (Illustration by Michael Blake after Taube 2004:13, figure 4)

Covarrubias, a noted Mexican artist and art historian, Coe observed that the were-jaguar, who was associated with the Rain God, often had maize sprouting from the V-shaped cleft in its head. The strong and frequent link between Rain God imagery and maize led Coe to suggest that one version of the were-jaguar could represent a distinct deity—the Maize God—in much the same way that there was overlapping imagery and symbolism among the representations of different gods among the later Maya, Zapotecs, Aztec, and many other cultures in Mesoamerica.[25] Later Coe and one of his graduate students, Peter David Joralemon, became the first scholars to recognize the significance of the corpus of Olmec depictions of maize in its various forms as belonging to an Olmec art tradition that depicted supernatural forces and deities linked to maize symbols.[26]

As mentioned in previous chapters we know that maize remains are associated with some of the earliest evidence of the Olmec occupation in Mexico. Many researchers have realized, however, that the frequency of maize in Olmec archaeological sites significantly increases around the beginning of the Middle Formative, about 3,000 years ago.[27] Karl Taube, an archaeologist and epigrapher who has done extensive research on Olmec and Maya symbolism, including documenting the representation of deities, argues that this is precisely the time when we see significantly more representation of deities of all kinds, including the Olmec Maize God. He notes that, although present, maize symbolism in general, and representations of the Maize God in particular, are rare in the Early Formative contexts at Olmec centers, such as San Lorenzo, and

FIGURE 9.8. Symbols representing maize, found on the Cascajal Block, which is thought to represent the earliest Olmec writing. (Illustration by Michael Blake, based on Rodríguez et al. 2006:1613, figure 5)

that it is not until the Middle Formative period that Maize God imagery becomes more frequent.[28] Many of these Middle Formative Maize God images are engraved on jade or other greenstone celts. These celts are symbolic and were never intended to function as practical tools—so for this reason they are called pseudo-celts. Artisans crafted them to represent supernatural forces—the greenstone celts symbolized a fresh ear of maize wrapped in its green husk. These objects are so clearly associated with maize that stylized images of celts representing the Maize God are incised into the surface of these pseudo-celts, creating a circular reference to maize (figure 9.7).

One of the earliest examples of Olmec signs that may be related to writing comes from a stone tablet recently discovered near San Lorenzo during a construction project. It was identified as a find of great significance by the Mexican archaeologists María del Carmen Rodríguez and Ponciano Ortíz. The tablet, called the Cascajal Block, is estimated to date to the transition between the Early and Middle Formative periods. It is made of serpentine and has a series of sixty-two glyph-like engravings constituting twenty-eight distinct signs (figure 9.8). The team who studied the tablet's signs recognized many of them as being similar to symbols defined by Joralemon decades earlier, but there were also many previously unknown signs. Of these, three are the same as maize symbols frequently found during the Middle Formative and later: "Signs 12, 17, and 27 show a thematic preoccupation with maize, or at least the ready use of such signs in the creation of a signary."[29]

Sign 1 is also closely associated with maize symbolism and is found in similar forms on some Early Formative ceramics at San Lorenzo in the Gulf coast region and the site of Cantón Corralito—an Olmec colony located in the Mazatán region on the Pacific coast, excavated in 2004 by David Cheetham (figure 9.9).[30] This use of symbols of maize is what we might expect if maize was becoming increasingly important toward the end of the Early Formative period, and it comes before we

FIGURE 9.9. Fragment of early Olmec Calzadas Carved pottery with an incised symbol possibly representing maize. This and similar Olmec style pottery were recovered by David Cheetham (2010b) during his excavations at the site of Cantón Corralito in the Mazatán region of Pacific coastal Chiapas, Mexico. (Photograph courtesy of David Cheetham)

see the explosion of Maize God imagery a few centuries later, during the Middle Formative.

Karl Taube points out that depictions of the Olmec Maize God are frequently associated with ballcourts and ballcourt symbolism. This is particularly interesting because the earliest Mesoamerican ballcourt was discovered at the site of Paso de la Amada, in the Soconusco region of Chiapas, Mexico—a site that has significant maize remains in the household deposits but that lacks any representations of a maize god or other maize symbolism as far as we know. But by the Middle Formative some ballcourts, such as the elaborate one at Teopantecuanitlán, in Guerrero, are laden with maize and Maize God imagery, including sculptures, architectural details, and mosaics.[31]

What does the Olmec Maize God look like? This has not been an easy question to answer because there is no direct ethnographic link between the earliest Olmec peoples and present-day indigenous peoples in Mesoamerica. However, scholars have been able to piece together well-recognized symbols found in more recent cultures and time periods—including the Maya and the Zapotec, for example—and trace many back to the corpus of Olmec motifs. Joralemon's work, published almost forty-five years ago, has been instrumental in helping scholars make these connections. The identification of the Maize God was made by recognizing a set of recurring motifs with a unique set of symbols. The main Middle Formative period representation is a face with the familiar Olmec "snarling" mouth—downturned with a protruding upper lip—and eyes that are often almond-shaped and appear "cross-

FIGURE 9.10. Depiction of the Olmec Maize God on a stone monument from the sunken court at the ceremonial center of Teopantecuanitlán in Guerrero, Mexico. (Illustration by Michael Blake after Taube 2004:97, figure 46a, based on Martínez Donjuán 1985, 1994)

eyed." The top of the deity's head usually has a V-shaped cleft from which emerges an ear of maize. There are several distinct representations of the Maize God, each of which can depict several stages of the growth of the plant. Some show the maize emerging from a dot below the cleft, thought to represent the seed or kernel. Others show the ear rising out of the cleft. Frequently the ear is shown surrounded by leaves, thought to represent the husk, while others show the exposed ear with bands around it, likely representing the kernels on the cob.

Joralemon initially hypothesized that other similar representations of deities, which lacked this particular set of characteristics, were distinct gods. Taube's analysis shows, however, that as many as four of the separate deities can now be considered to be representations of different stages, or aspects, of the growth cycle of maize: from planting the seed to sprouting, growing, maturing, and ripening.[32] The Maize God sculpture from Teopantecuanitlán is a vivid example of this, with several different aspects of the maize growth cycle represented in one image (figure 9.10).

Versions of the Maize God continue to be represented in later periods and extend beyond the Olmec heartland to many other regions of Mesoamerica. The evolution of the forms of deities from Olmec antecedents was noted by Covarrubias in the 1950s and picked up on by many other scholars. It is fascinating to see the visual and symbolic connections between the earliest Olmec representations of the Maize God and later versions portrayed in Classic period Zapotec and Maya cultures, and even subsequent Postclassic Mexican cultures.[33]

FIGURE 9.11. Depiction of the Olmec Maize God
(labeled by the researchers who discovered it as
Individual 9) in a mural painting on the north wall of a
temple at the Late Preclassic Maya site of San Bartolo,
located in northeastern Petén Region of Guatemala.
(Illustration by Michael Blake, after Saturno et al.
2005:27, figure 20a, based on a drawing by Heather
Hurst)

One of the most exciting archaeological discoveries in Mesoamerica
in the last decade was made by William Saturno, who led an expedition
to the northern Petén region of Guatemala, where he and his team of
researchers found the earliest known Maya polychrome murals—beautifully preserved in an ancient pyramid at the San Bartolo site.[34] Tucked
into the far northeast corner of the lowland tropical jungle near the
borders of Guatemala, Mexico, and Belize, the site had been partly
looted by tomb robbers just before Saturno and his team arrived. Sal-
vaging what they could from the looted tunnels, the archaeologists
found Maya hieroglyphic writing on the insides of the temple walls—
which turned out to be centuries earlier than the much better known
Classic period Maya texts from nearby sites, such as Tikal. Because the
site is Late Preclassic in age, it falls between the Early and Middle Preclassic Olmec cultures, which we've just been discussing, and the Classic period. While these murals are similar in many respects to Classic
Maya painting traditions found elsewhere in the Maya lowlands, their
style shows clear connections to earlier Olmec styles and iconographic
depictions—including early Maya maize god representations that
are strikingly similar to depictions of the Olmec Maize God (figure
9.11). The layout and content of the San Bartolo murals anticipate
recurring maize symbolism in later periods, including references to
events in the epic Popol Vuh and the four world directions and their
color associations.[35]

SOUTH AND NORTH AMERICA

Images of maize are not restricted to Mesoamerica, although that is where such depictions first appear. To the south we see spectacularly detailed depictions of maize on Moche pottery from the north coast of Peru, dating to about 1,900 to 1,200 years ago. Mary Eubanks has written about maize in Moche art and compared it with contemporary depictions of maize on Classic period Zapotec urns from Oaxaca, Mexico.[36] The Moche people, she argues, portrayed husked ears of maize so accurately that one can tell the variety—some of which appear to her to have been imported from Mexico. Likewise, there are Zapotec ceramic urns with depictions of maize that appears to be derived from South American varieties. Unfortunately there are too few actual preserved maize ears with kernels intact to be certain what maize varieties were present in Oaxaca during the Classic period, so one must have a great deal of confidence in the realism of the maize depictions on these vessels to be certain that a direct exchange of maize varieties between these places took place.

One of the differences between Mesoamerican representations of maize and the Maize God and Moche maize imagery is that, in Moche culture, maize was most often linked to a god through action more than it being an essential aspect of the deity itself. The Moche fanged deity, which is also called the "Decapitator" and frequently represented in scenes of sacrifice, is seen holding maize in one hand and manioc tubers in the other (figure 9.12).[37] This deity seems to be rooted in the earlier Chavín jaguar deity—a powerful image that reverberates through all subsequent Andean cultures right up to the time of the Spanish Conquest.

Renowned art historian Elizabeth Benson has noted that Moche pots with images of the fanged deity include more depictions of maize than any other type of food—with the possible exception of manioc (figure 9.13). Although we don't know exactly what such pots were used for, they are most commonly found in association with burials, suggesting that they may have contained ritual offerings of *chicha* to accompany the deceased in the afterlife.[38]

One of the most spectacular Andean images of maize comes from a first-hand description by Cieza de León, one of the conquistadors who accompanied Francisco Pizarro in the conquest of the Incas. Many years after the conquest Cieza de León recounted seeing the splendor of the Qorikancha—the temple to Inti, the Sun God, which was centered in the heart of the city of Cuzco, at the intersection of the main axes that defined the entire Inca Empire. One of the four gold-covered houses in

FIGURE 9.12. Moche maize god depicted on a moldmade pottery vessel from the north coast of Peru. This image shows what are probably tubers in the right hand and a maize plant in the left. (Illustration by Michael Blake, based on Berezkin 1980:28, figure 6)

one of the rooms of the Qorikancha contained "the figure of the sun, very large and made of gold, very ingeniously worked, and enriched with many precious stones. . . . They had also a garden, the clods of which were made of pieces of fine gold, and it was artificially sown with golden maize, the stalks, as well as the leaves and cobs, being of that metal."[39]

CHICHA

The imagery of the Inca temple Qorikancha gives a clear indication of the symbolic importance of maize in the Andean region of South America—an importance that extends from at least the Initial period, about 3,500 years ago, right up to the present day. As we touched on earlier, maize had been—likely from its earliest introduction—important in the Andean region because of its role in making *chicha*. *Chicha* was especially significant for the Wari and Tiwanaku peoples of highland Peru and Bolivia. The Wari are renowned for their large *chicha* brewing facilities. One particularly well-preserved example of a *chicha* brewery was discovered at the ancient site of Cerro Baúl, a nearly impenetrable mesa-top settlement located at the boundary between the Wari and Tiwanaku empires. During excavations at this site in the early 2000s, Michael Moseley and Ryan Williams came across a large set of rooms dedicated

FIGURE 9.13. Pot showing the Moche fanged deity's head emerging among four ears of maize, modeled in realistic relief, which combine to form a peak, behind which is a large spout. This vessel was used as a mortuary offering, dating to the Moche IV period (ca. 450–550 CE) (Benson 1997:126). It is one of thousands of such vessels and other artifacts that can be seen in the vast open storage collections at the famous Museo Arqueológico Rafael Larco Herrera in Lima, Peru. (Photograph by Michael Blake, of a Moche pot found in the Museo Larco collection, catalogue number ML006622; see www.museolarco.org/catalogo/ficha.php?id=6622 for other images of this vessel)

to grinding maize, cooking maize mash in large ceramic vessels, and fermenting the resulting brew in pots to make *chicha*. Their excavations showed that the brewery, and indeed probably the whole settlement, had been rapidly abandoned about a thousand years ago. It appears that just before leaving the site, the garrison that was stationed at the hilltop fortress consumed all of the remaining *chicha,* drinking it from special ceramic beer mugs—ceremonial cups called *keros*—some of which were decorated with the image of the Front-Facing God (figure 9.14). They then tossed the *keros* on the floor of the fermentation patio and, as a final act of departure, burned down the whole complex before leaving it forever.[40] This type of *kero,* found widely at Middle Horizon period sites throughout the Andean region, is often distinctively decorated with images of the Front-Facing God—reminiscent in many ways of the Moche people's Maize God and harkening back to some of the imagery of the Staff God found during the Early Horizon, many centuries before.

This tradition of *chicha* brewing and ceremonial drinking is well documented for the Inca and their descendants, both ethnohistorically and ethnographically. The Inca, like the Wari and Tiwanaku before

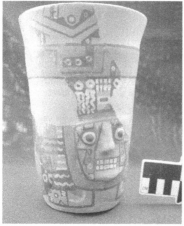

FIGURE 9.14. Two *chicha* drinking cups, or *keros*, each showing the Front
Facing God. These were recovered in excavations at the Middle Horizon
Wari site of Cerro Baúl, a hilltop garrison located in the lower Moquegua
River valley of southern Peru. Excavators Patrick Ryan Williams and
Michael Moseley and their team found dozens of such *keros* that had been
broken and left on the floors of drinking rooms—likely discarded when the
site was abandoned. (Photograph courtesy of Patrick Ryan Williams)

them, had a complex and well-integrated agricultural system and grew
vast amounts of foods. Their two main food staples were potatoes and
maize. A freeze-drying process was used to preserve potatoes for transport and storage on a massive scale, and they were an essential component
of the financing of imperial construction projects and military expansion. Likewise, maize was grown, harvested, and stockpiled in storage
facilities and warehouses interspersed throughout the vast Inca political
domain. Whereas tubers such as potatoes and seed crops such as beans
and quinoa were used as foods in daily meals, maize may have been
used primarily for brewing *chicha*. All social interactions were characterized by reciprocity, and maize beer consumption was one of the cornerstones of every social and ritual event, large or small. In the Andes,
maize was a food, but when it was converted to *chicha* it became much
more important than a simple foodstuff—it was transformed into a
drink from and for the gods.

In sharp contrast to the Andean region of South America, in North
America maize was used most often as a food and seldom for fermented
beverages, at least prior to the arrival of Europeans. But maize was
symbolically important in some regions—particularly where it had been

grown and relied on for the longest periods of time. Maize moved north out of what is modern-day Mexico, brought to the southwestern United States at least 4,000 years ago, either by migrating farmers, trade from community to community, or perhaps a combination of the two. Maize farmers planted it in the upland zones of New Mexico and in the low, hot river basins of southern Arizona. Over the next thousand years maize became increasingly important, but there is no evidence of maize symbols from these early periods. Most of what we know about the plant's earliest use in this region comes from the charred remains of kernels, cupules, and occasional cob fragments.

One of the most recent syntheses of maize symbolism in the US Southwest comes to us from the work of Karl Taube, who is in the unique position of being able to compare developments in the that region with those in Mesoamerica. The Puebloan peoples of the greater southwest, which extends from the states of Chihuahua and Sonora in Mexico to Utah and Colorado in the United States, have many different traditions and histories—but maize runs through them all. Most present-day Puebloan peoples think of maize as sacred and believe that humans, literally and symbolically, are maize and that maize is human. At the most general level, ears of maize, and the particular colors of the kernels, are intimately linked to the four world directions—making maize a fundamental aspect of Puebloan cosmology.[41] The cosmological significance of the plant is symbolized most clearly by the use of maize "fetishes." These are sacred bundles consisting of an ear of maize—generally a perfect ear with no missing grains—and are usually wrapped in feathers. Taube details the physical similarities and symbolic connections between Puebloan maize ear fetishes and their Mesoamerican counterparts. In the southwest the feathers used to wrap the sacred ear of maize often come from macaws— the feathers of which, or even live birds, had to be brought from far to the south as part of a very long-distance exchange network. One of the most fascinating aspects of these sacred fetishes is that they are in most respects identical to the maize fetish tradition that first appeared 2,500 years earlier among the Olmec of southern Mesoamerica. Maize fetishes— complete with macaw or quetzal feathers—are especially common in the corpus of Olmec maize symbols dating to the Middle Formative period. Archaeologists have found surprisingly similar images at two Pueblo IV period village sites dating from 1350 to 1600 CE: Awatovi in Arizona and Pottery Mound in New Mexico (figure 9.15).[42]

This means either that such fetishes had been used by ancestral Puebloan peoples for 2,500 years or that they were adopted more recently

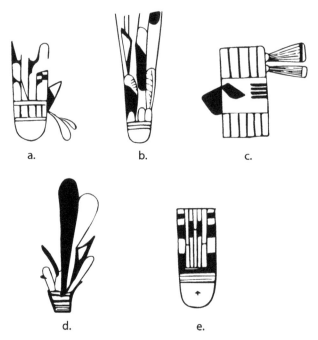

FIGURE 9.15. Pueblo IV period maize fetishes represented in murals from the Awatovi Pueblo site in Arizona (a–c) and the Pottery Mound site in New Mexico (d-e). (Illustration by Michael Blake after Taube 2000:324, figure 27d–h)

from neighboring groups with direct connections to populations in Central Mexico, where they were also used in sacred rituals, and in some cases still are. Either way, this symbolic connection is more than simply coincidental; it must be linked to the underlying principles emphasizing the role of maize in the creation of the Puebloan cosmos.

The maize symbolism connections between the Puebloan peoples and Mesoamerican cultures go even further. Taube shows that special celts, called *chamahiya,* symbolize lightning and rain, and, like the maize fetishes, are intimately linked to the world directions and the axis mundi.[43] As in the case of the ceremonial greenstone celts used and depicted in Olmec ritual contexts, these Puebloan *chamahiya* celts were never functional in the sense that they could be used to chop or carve. Their functions were much more important. They were deployed on ceremonial occasions to ensure the coming of the forces of nature that secured the survival of the Puebloan world: the rain and lightning that brought agricultural fertility and productivity.

One of the earliest depictions of *chamahiya* celts that link maize and lightning comes from the Three Rivers area of southern New Mexico and dates to between about 1000 to 1400 CE (figure 9.16). It shows lightning, in the form of a maize plant with three celt-shaped ears, descending from a cloud, upon which is perched a bird. This image links maize with the agricultural productivity of clouds and rain and with celts and lightning. In other examples of *chamahiya* representations in the US Southwest, the same cloud shape is often shown as one of four such symbols which are aligned, much like a compass rose, to indicate the four world directions. And, to top it all off, the bird sitting at the apex of the cloud links the symbolism of feathers to maize, as we saw earlier with the maize fetishes.[44]

Maize symbolism in the US Southeast—and extending to the Midwest and Northeast and to southeastern Canada—is much less common than in the US Southwest. Peoples who moved north, from the mouth of the Mississippi upriver to all of its tributaries and as far as the Great Lakes, practiced maize farming as part of a large and complex agricultural suite. Some of the earliest dates for maize in this region come from the Holding site near Cahokia in Illinois, just southeast of St. Louis, Missouri. Directly dated samples of maize place the adoption of the plant at slightly earlier than 2,000 years ago. Much later, by the time of the Mississippian culture, a period of large villages and vast networks of people who participated in the Southeastern Ceremonial Complex from about 1200 to 1650 CE, there is evidence for significantly increased reliance on maize agriculture. It seems that people adopted maize as a food staple, even preparing it using nixtamalization to produce hominy, but they did not accord it the same sacred status as did their neighbors to the Southwest, and there is very little evidence for specific maize iconography or general maize symbolism.

There are a great many symbols in the Southeastern Ceremonial Complex, but none are specifically concerned with maize in the way we saw in Mesoamerica and the US Southwest. The Mississippians and their neighbors in the Midwest and Northeast treated maize as more of a utilitarian food source—albeit an extremely important one. For them the parallel is much closer to what we see in much of South America, where maize was certainly significant—especially in the Andean region, with its focus on *chicha* production—but it was not necessarily sacred. Although the Mississippians didn't make their maize into *chicha,* they, like many South American peoples, grew it and relied on it heavily as a food source, although only as one of many domesticated foods—most of which originated in eastern and southeastern North America.[45]

FIGURE 9.16. A maize *chamahiya*, or lightning celt, depicted on a petroglyph (natural stone carving) from the Three Rivers region of southern New Mexico. Karle Taube argues that this image portrays a maize plant with three celt-shaped ears descending from a "directional" cloud and a bird. The maize plant itself may represent a bolt of lightening. This echoes the associations of maize, celts, lightning, the world axis, and world directions, which are seen in depictions found throughout Mesoamerica. (Illustration by Michael Blake, after Taube 2000:326, figure 28c)

One exception to the general lack of symbolic representation of maize in the eastern United States is the practice of "Green Corn" rituals. These were ceremonies linked to the first appearance of new maize during the summer, and in the Mississippian region they often included the amplification and rebuilding of earthen platform mounds. Vernon Knight describes the mounds as being "quadrilateral, flat-topped," and "probably connoting the southeastern cosmological concept of earth as a discrete, flat-surfaced entity oriented to four world quarters."[46] This signifies the symbolic importance of maize as well as its economic and nutritional status, and it is certainly reminiscent of the connections the Mesoamericans made between maize and the four world directions and the axis mundi. The Green Corn Ceremony was more than just an occasion to acknowledge the bounty of the new harvest; according to some anthropologists, rebuilding ceremonial mounds was a way to renew and purify the society as a whole.[47]

GROWING, EATING, THINKING, AND BEING MAIZE

From the Great Lakes in North America to the highlands and lowlands of South America, wherever maize was grown and eaten it was also revered. In some cases this reverence was no more or less than that accorded other major and minor foods that made up people's daily fare.

In others the importance of maize went beyond the quotidian and dietary, standing for the fundamental organization of a community's worldview and structuring the way people envisioned the very cosmos and its creation. For the Quiché and other Maya groups of highland Guatemala, this essential aspect of maize is reflected in the story of the Popol Vuh, a text from the very early colonial period that recounts the Quiché origins and cosmology. The part of the history that details the creation of the first humans states explicitly that they are made of maize: the forefathers, Tepeu and Gucumatz, "began to talk about the creation and the making of our first mother and father; of yellow corn and of white corn they made their flesh; of cornmeal dough they made the arms and the legs of man. Only dough of corn meal went into the flesh of our first fathers, the four men, who were created."[48] Many of the images and descriptions of events in the Popol Vuh can be traced back to the Classic period, where they appear in paintings on pottery and imagery carved in stone. These ancient images, in turn, echo similar images and depictions in the Early Classic and Preclassic periods, going back to Olmec times.

For ancient Mesoamericans the symbolic and dietary importance of maize seemed interlinked. Maize took on a visual symbolic role, represented in material expressions and exquisite artistry, at approximately the same time that there is increasing evidence for its use as a major food staple rather than just a supplement. The ancient Mesoamericans' increasing consumption of maize, beginning around the time of the earliest settled villages, shows up in the chemical signatures of their bones. This process accelerates over several centuries so that the early Olmecs of San Lorenzo and neighboring communities in the Gulf Coast region of Mexico had a suite of symbols that refer to maize by about 3,000 years ago. Even more telling is that some of these symbols are directly linked to a god of maize only a few centuries later.

This trajectory of symbolic promotion—whereby maize imagery links to and expresses the increasing social, political, and economic significance of what was once just one of many important foods—is not surprising. Foods for feasting and foods for daily living are different from one another. Some foods are special and others are mundane. For many peoples, foods are also very closely connected to identity. For Mesoamericans, as expressed so eloquently in the Popol Vuh, food—in this case maize—became people. This was also true of the ancient Puebloan peoples of the southwestern United States, who either brought maize with them as they occupied the mesas and valleys of their vast

landscape or adopted maize as a nutritionally important and symbolically loaded import from their Mesoamerican neighbors to the south.

To the north, beyond the US Southwest, maize did not take on the
same symbolic role, even though it eventually became an important staple; along with beans and squash, the three foods were known as the
famous "three sisters" of the northeastern peoples of North America,
such as the Iroquois. To the south, beyond the southern bounds of Mesoamerica, there are several instances where maize was present from a
very early date but where the symbolism associated with maize is not
nearly as elaborate as it was in Mexico and the Maya region. For example, archeologists working in parts of Central America that were sup-
posedly settled by Nahuatl-speaking migrants from the north seldom
find much evidence of maize or maize cooking utensils.[49] This seems
puzzling since we associate maize so closely and for such a long time
period with the populations of Mexico. Why would these migrating
peoples, or even the local populations themselves, not be equally enam-
ored with maize? One possibility is that they were much more wedded
to the production of manioc—an extremely productive food source that
is easily planted, stored, and harvested. Casts of manioc roots growing
in fields and stored in people's abandoned houses were preserved under
a thick layer of volcanic ash that covered the Maya site of Joya de Cerén
in central El Salvador around 600 CE.[50] If manioc was in use in ancient
times throughout Central America, then its importance may have
eclipsed that of maize—or perhaps lessened its appeal as a staple food.
This is similar to the pattern we see continuing south into the Andean
regions as well as in Amazonia, where root and tuber crops may have
had the upper hand, compared with maize, because this is where they
were first domesticated and used as staple foods. As a further point in
this hypothesis, manioc was traditionally prepared like maize in many
regions of South America where it is grown: it can be cooked on ceramic
griddles in a similar fashion as maize tortillas, and it can also be cooked
and fermented to make a type of manioc *chicha*.

In the sites that I've had the privilege of excavating and studying,
from Pueblo ruins in the Mimbres Valley of southwest New Mexico to
Postclassic Canajasté in the Maya highlands of Chiapas to Early Formative sites in the Soconusco region along the Pacific Coast, one thing has
been constant: people's interest in and reliance on maize. Though these
few sites give only a snapshot of the range of dependence on and importance of maize in ancient political and ritual economies, it is clear that
maize use was never constant. The ways people grew their precious

food, prepared it, thanked it, and shared it with their gods varied across the vast expanse of the continent and through the centuries and millennia. In this way little has changed. We see today, all around us, the transformation of both maize and people as it has spread to almost every place on the planet where crops are grown—and even where they aren't, maize still manages to find a way there.

Notes

INTRODUCTION

1. *Vancouver Sun,* January 30, 2007; Reuters Online, May 22, 2012; Reuters Africa Online, March 23, 2012.

2. FAOSTAT, http://faostat.fao.org/site/567/DesktopDefault.aspx?PageID = 567#ancor.

3. In North America "corn" is the most common name in English for *Zea mays* ssp. *mays.* In Spanish the most common term is *maíz.* I will use the English-language term "maize" throughout this book because it unambiguously identifies *Zea mays* without the potential confusion that arises from the word "corn," which in the United Kingdom refers more generally to grain of all kinds.

4. Willis 2002; Papas et al. 2010.

5. Mangelsdorf et al. 1964; Sanders and Marino 1970.

6. Blake 2010:260.

7. Clark 1994.

8. Ceja Tenorio 1985; Clark 1994.

9. Feddema 1993.

10. Blake et al. 1995

11. Chisholm and Matson 1991.

12. Blake et al. 1992a.

13. In 2000 the botanist Hugh Iltis published a paper in the journal *Economic Botany* outlining the sweet stalk hypothesis and proposing that maize may have been domesticated by early farmers who were interested in the sugary juice that could be easily obtained from the stalks of the teosinte plant—much like sugarcane and other succulent grasses. John Smalley and I further developed this hypothesis in 2003 and accumulated ethnographic, archaeological, and isotopic evidence to help evaluate it.

14. Goldemberg et al. 2004.

15. Bruman 2000.
16. UNESCO 2010.

I. THE ARCHAEOLOGY OF MAIZE

1. Wilkes 1967; Mangelsdorf 1974.

2. Flannery 1986c.

3. Throughout this book the ages of settlements, samples, and events will be presented in several formats. One of the most common usages among archaeologists is the abbreviation "**BP**," which stands for "years before present." By convention we take "present" to mean 1950—the date after which radioactivity in the atmosphere began to rise dramatically as a consequence of post–World War II atomic bomb testing. The abbreviation "**cal BP**" stands for "calibrated years before present"—that is, a radiocarbon (^{14}C) age estimate that has been "calibrated" or corrected, by matching it to age estimates derived from a massive sample of radiocarbon-dated tree rings of known calendric age. Matching raw radiocarbon dates to the tree-ring date curve allows ^{14}C dates to be converted from "radiocarbon years," which often give ages that are decades or even centuries older or younger than their corresponding tree-ring calibration dates. The result is more accurate dating (Bronk Ramsey 2008; Bronk Ramsey 2014). Finally, archaeologists will occasionally refer to dates in the calendric system using the updated notations **BCE** ("before the Common Era") and **CE** ("Common Era"). These terms are increasingly used in current literature to replace the Christian dating notations BC ("Before Christ") and AD ("Anno Domini").

4. Domestication is the process of transforming a species of plant or animal from a wild state to one in which it is accustomed or adapted to living with and under the care of humans.

5. The same is true for the domestication of animals—whether for food and materials (e.g., cattle, pigs, and chickens) or service (e.g., dogs, horses, and cats).

6. Michael Pollan (2001, 2006) presented an elegant version of this argument in his award-winning, popular books *The Botany of Desire* and *The Omnivore's Dilemma*.

7. Mangelsdorf 1974.

8. Beadle 1980.

9. Wilkes 1967; Doebley 2004.

10. van Heerwaarden et al. 2011.

11. Fukunaga et al. 2005:2241–2242.

12. Botanists have divided both *parviglumis* and *mexicana* into five geographic sub-groupings, or "races." For *parviglumis* the groups are: Central Balsas, Eastern Balsas, Jalisco, Oaxaca, and Southern Guerrero. *Mexicana*'s groups are: Central Plateau, Chalco, Durango, Nobogame, and Puebla. Fukunaga et al. 2005; Wilkes 1967.

13. Fukunaga et al. 2005.

14. Iltis and Doebley 1980.

15. Iltis 2006:26.

16. Matsuoka et al. 2002.

17. Flannery 1986a; Piperno and Flannery 2001.
18. van Heerwaarden et al. 2011.
19. Pollard 1997.
20. Benz 2000.
21. Benz 2001.
22. Ammerman and Cavalli-Sforza 1984; Bellwood 2005.
23. Piperno et al. 2009; Ranere et al. 2009
24. Blake 2006; Blake et al. 2012.
25. Long et al. 1989.
26. Smalley and Blake 2003; Iltis 2000
27. Bruman 2000.
28. Benz et al. 2006; Mangelsdorf 1974.
29. Mangelsdorf 1986.
30. Smith 2006:17–19.
31. Diehl 2005; Mabry 2008.
32. Benz et al. 2006.
33. Burger and van der Merwe 1990.
34. Staller 2006, 2010.
35. McGovern 2009:206–209.

2. THE PLACE OF MAIZE IN (AGRI)CULTURAL ORIGIN STORIES

Epigraph: Smith 1995:3.

1. In all societies the substances that are consumed are classified in many different ways, and most of these classification schemes have moral overtones and social implications. Some foods and drinks are thought to be disgusting while others are considered to be pure ambrosia. This doesn't just apply to those substances that contain mind-altering chemicals (Wenk 2010). A few years ago I invited some colleagues who were visiting from Europe over to my house for a North-American style barbeque. While they thought the dinner was delicious, especially the wild-caught sockeye salmon, they declined to eat the corn on the cob, sheepishly admitting that, where they came from, it was considered food for livestock and not really something that people would consume.

2. Trigger 1989; Pluciennik 2001; Douglas 1984; Bourdieu 1984; Lévi-Strauss 1969.

3. Estimates vary depending on whether one is looking at archaeological evidence, which suggests that behaviorally modern humans have been around for about 50,000 years, or fossil and DNA evidence, which suggests that anatomically modern humans have been around for 200,000 years (Mellars 2006).

4. Pluciennik 2001.
5. Wilson 1988.
6. Tylor 1865.
7. Smith 1995:3.
8. Pluciennik 2001.
9. Kroeber 1933:414.
10. Marx 1965; Engels 1972.

11. Morgan 1877:22.

12. Smith 1812 (vol. 1):317.

13. Smith 1812 (vol. 2):57.

14. Smith 1812 (vol. 2):358.

15. Smith 1812 (vol. 2):351.

16. Pattberg 2007; Staller 2010:80–81.

17. Psalm 104:14 (English Standard Version Anglicised).

18. Dolores Piperno and Deborah Pearsall's book *The Origins of Agriculture in the Lowland Neotropics* (1998) is a notable exception. Specialists in the agriculture of the tropical regions of the Americas, they examine the processes of agricultural origins and food production systems that were, in many cases, very different from those that characterized the initial stages of agriculture in the Near East.

19. Bellwood and Renfrew 2003; Bellwood 2005.

20. New studies of the genetic markers identified in the ancient DNA of European Neolithic hunters and farmers—which was recovered from cemetery sites in various regions of Europe and dated to about 7,000 to 8,000 years old—show the complex process of gene flow in these early populations (Lazaridis et al. 2014). As more and more ancient individuals from Europe, the Near East, and Central Asia have their DNA sequenced, our knowledge of the complex relationships between ancestral European hunter-fisher-gatherers and in-migrating farmers from the Near East will become richer and more complex.

21. McCann 2005:23.

22. Díaz del Castillo 1963.

23. Díaz del Castillo 1963:14.

24. Díaz del Castillo 1963:99. Cortés's expedition was provisioned in Cuba where cassava, rather than maize, was the main food staple.

25. Díaz del Castillo 1963:100. Cortés was so certain that he and his army would be able to live off the land that he eventually scuttled his ships to prevent any thought of a return trip. However, the conquest could not have proceeded without the plentiful supply of maize and other local foods.

26. Díaz del Castillo 1963:174.

27. Díaz del Castillo 1912:294–295.

28. For example, there has been a long-running debate about the significance of agriculture, and maize in particular, in the subsistence economies of early complex societies in the Andean region. See Moseley 1975, 1992; Wilson 1981; Haas and Creamer 2006; Raymond 1981; Quilter and Stocker 1983; Shady 2006; Shady Solís and Leyva 2003 for detailed arguments over the past thirty years.

29. Willey 1966; 1971.

30. Tedlock 1985.

31. Girard 1979.

3. OLD PUZZLES AND NEW QUESTIONS ABOUT MAIZE'S ORIGINS AND SPREAD

Epigraph: Mangelsdorf 1974:2.

1. Recent varieties of supersweet corn, commonly found in North American marketplaces during the past decade, are a result of hybrid breeding of varieties

of sweet corn that have particular mutations of the sugary locus (*su*), which controls sugar content and quality, in particular the genes sugary enhanced (*se*) and shrunken-2 (*sh₂*) (Schultheis 1998).

2. Mangelsdorf 1974:21.

3. Wills et al. (2013) provide a fascinating recent discussion of the genetics of teosinte's ear structure and the changes that the plant underwent during domestication—which include the development of specific genes that, in maize, prevent the development of multiple spikes on one ear.

4. A maize ear, however, is not simply several teosinte ears merged together. Instead, the maize ear is the result of one dominant spike growing at the expense of the subordinate spikes. This is an example of what botanists call apical dominance, when the primary flower takes all the nutrients and the secondary and tertiary flowers are stunted or aborted. In maize, genes controlling this process of apical dominance interacted with other genes, eventually leading to multiple rows of paired spikelets (polystichy) rather than just one distichous row of spikelets (Iltis 2006; Bommert et al. 2013).

5. Both Garrison Wilkes (1967:63–70) and Paul Mangelsdorf (1974:21–35) describe and illustrate these differences between teosinte and maize in their authoritative books on the topic.

6. Wilkes 1967:63.

7. Paul Mangelsdorf (1974:xi) made the following statement about why it is important to study the origins of maize: "In recent years when it has become apparent that the pressure of an explosively expanding world population is creating worldwide food problems, I have sometimes argued that botanists should know as much about the world's principal food plants, of which corn is one, as we know about some of the destructive agents of the world, as the medical profession, for example, knows about the principal human diseases, and as aeronautical engineers know about the world's principal bombers and guided missiles."

8. James W. Cameron is listed in the preface to *Corn: Its origin Evolution and Improvement* (Mangelsdorf 1974) as one of a team of graduate students who contributed enormously to the understanding of maize in the twentieth century.

9. "The tripartite hypothesis that cultivated maize had its origin in South America as a single gene mutation from a wild form of pod-corn; that Euchlaena [teosinte] is a recent product of the natural hybridization of Zea and Tripsacum which occurred when the two genera were brought together in Central America, and that new types of maize originating from this cross comprise the majority of North American varieties, is in accord, we believe, with all the known facts" (Mangelsdorf and Reeves 1938:311, 1939). This statement led to a forty-five-year-long quest for genetic and archaeological evidence that maize: (1) had at least one separate origin in South America, (2) had a wild extinct ancestor—a type of pod-corn—and (3) was itself the ancestor of teosinte, rather than the reverse.

10. Beadle, along with other botanists at the time, recognized the significance of the nineteenth-century German botanist Paul Ascherson's conclusions that teosinte was the closest wild ancestor of maize.

11. Beadle 1939:246.

12. See John Doebley's (2001) thorough summary of this debate and discussion of Beadle's prescient observations concerning the genetic changes that teosinte underwent as it was transformed into maize.

13. Doebley 2001; Horowitz 1990.

14. Mangelsdorf et al. 1964:541.

15. Doebley 2001, 2004; Iltis 2006.

16. Iltis 2006:24.

17. Garrison Wilkes's doctoral dissertation (1967) surveyed and described all the varieties of teosinte, and it still serves as the definitive description of the plant, its habits, and habitats.

18. In addition to Balsas teosinte, Wilkes described Chalco teosinte (*Zea mays* ssp. *mexicana*) (Wilkes 1967).

19. Iltis and Doebley 1980.

20. Beadle 1980.

21. Doebley et al. 1984. Emshwiller (2006) provides a clearly written summary of the history of these genetic studies and their significance for understanding maize domestication (see also Doebley 1990).

22. Doebley 2001:491; Mangelsdorf 1983, 1986.

23. MacNeish and Eubanks 2000; Eubanks 2001a, 2001b. See Bennetzen et al. (2001) for a critique of this perspective based on the genetic evidence available up to that date. Chapter 8 reviews the current genetic information between 2001 and the present.

24. Mangelsdorf 1974:120.

25. Ammerman and Cavalli-Sforza 1984.

26. Rowley-Conwy 2011:S434.

27. Rowley-Conwy 2011:S434.

28. Hill 2001; Matson 1991, 2003; LeBlanc et al. 2007.

29. Merrill et al. 2009.

30. Carl O. Sauer (1889–1975) was one of the twentieth century's most influential geographers. His ideas about the causes and consequences of plant and animal domestication and their movement—best captured in his 1952 book *Agricultural Origins and Dispersals*—continue to resonate among archaeologists, anthropologists, geographers, and botanists alike.

31. Some authors are explicit about preferring one theory for agricultural origins and development over another. For example, in their book *The Origins of Agriculture in the Lowland Neotropics,* Dolores Piperno and Deborah Pearsall state: "Behavioral ecology seems to us to be the most appropriate way to explain the transition from human foraging to food production" (Piperno and Pearsall 1998:16). Others take a more eclectic approach, and in recent years scholars have been focusing more on the regional histories of plant use within specific social and geographical contexts rather than relying too heavily on general explanatory processes. This bears directly on the connections one sees between food production and social and political organization. For example, in discussing complex developments in the Amazon, Neil Whitehead et al. (2010:91) emphasize that "the earlier relegation of Amazonian societies to the 'Tropical Forest' culture type as more having resulted from intellectual biases

about what constitutes 'civilization' as well as the over-use of Old World archaeological criteria for measuring socio-cultural complexity through such traits as urban architectural forms, writing or glyphics. Instead, scholars are now developing a better appreciation of the alternative meanings of complexity in the Amazonian context." To a great extent this includes many new discoveries of the vast and ancient network of communities that so extensively modified Amazonian landscapes with their practices of food production and social reproduction (Heckenberger 2005; Erickson and Balée 2006).

32. Flannery 1986c:4.

33. Flannery 1986a.

34. Hayden 2009; 1990.

35. Matson 2003; Coltrain et al. 2007.

36. Marcus and Flannery 1996:71–72; Kirkby 1973.

4. TIMING IS EVERYTHING

1. Darwin 1846:131–132.

2. Darwin 1846:135.

3. Uncharred plant remains, including those of maize, seldom preserve in the open because the biological processes of decay act quickly to convert the energy stored in plant cells into food for thousands of microorganisms.

4. Libby et al. 1949.

5. Libby and his team first tested this method with two wood samples taken from Egyptian tombs of known age (approximately 4,600 years old). They found that the reduced rate of beta particle emissions from the two samples corresponded well with their known age (Libby et al. 1949). In their very first announcements of the radiocarbon dating method, Libby and his team used a half-life of 5720±47 years. By early 1951 they recalculated the half-life to 5568±30 years, using what they thought were more accurate measurements. Ironically, today the standard half-life in use is 5730±40 years (almost identical to Libby's original estimate from the 1940s). An original "Libby" date calculated using the 5568 half-life should be converted by multiplying it by 1.03 to obtain the "correct" radiocarbon age, according to the current standard. Where possible, the dates mentioned here have been converted to the new standard half-life and will therefore be found as slightly younger dates in the original publications. See Bowman (1990) and Taylor and Bar-Yosef (2014) for a comprehensive discussion and explanation of radiocarbon dating methods and new developments.

6. Mangelsdorf 1974:147; Mangelsdorf and Smith 1949.

7. Arnold and Libby 1951:116. The calibrated dates are somewhat earlier: No. 167 is 2000 to 1400 cal BP, while No. 171 is 2750 to 2050 cal BP.

8. The original Libby dates are 5605±209 and 1610±200, respectively.

9. Mangelsdorf 1974:149.

10. Libby 1952:679–680. The original date is 4445±180.

11. Mangelsdorf 1974:153.

12. MacNeish 1958.

13. MacNeish and Peterson 1962; Flannery and Marcus 2001; Mangelsdorf 1974:165.

14. The five volumes, under the general editorship of Scotty MacNeish. are titled *The Prehistory of Tehuacán Valley,* Vol. 1 (Byers 1967), Vol. 2 (MacNeish et al. 1967), Vol. 3 (MacNeish et al. 1970), Vol. 4 (Johnson 1972), and Vol. 5 (MacNeish et al. 1972).

15. Mangelsdorf 1974:167.

16. As amazing as this huge sample is, even more maize remains were recovered from Tularosa Cave, located in New Mexico, in 1950. Paul Martin of the Field Museum of Natural History found more than 33,000 cobs from deposits in that site. Some of these were directly dated by Libby in his initial radiocarbon dating program at the University of Chicago (Libby 1951:293–294).

17. Mangelsdorf 1974:168–169.

18. Mangelsdorf 1974:182, figure 15.25.

19. I first met Erle Nelson four years prior to his seminal work on AMS dating during an archaeology field school that I took part in as a first-year undergraduate student. On a visit to our excavations along the Fraser River, in the Interior of British Columbia, Erle demonstrated to our field crew how another new method of dating, called thermoluminescence dating, worked. Although this method doesn't work on maize or other organic materials, it does work on ancient artifacts, such as pottery and some types of stone.

20. Bennett et al. 1977; Nelson et al. 1977.

21. Wills 1988. All calibrated ^{14}C dates in this and the remaining chapters are calculated using the online program Oxcal 4.2, developed and maintained by Christopher Bronk Ramsey and his colleagues at Oxford University (Bronk Ramsey 2014). The ^{14}C dates reported here are given as calibrated years before present (by convention, "present" is taken to be 1950): "cal BP." Where I give a range of dates, such as 2795–3555 cal BP, the range represents a statistical probability of 95.4 percent (two standard deviations) that the sample's true age falls within this time span. In most cases I round up the age estimates to the nearest five-year interval. To make the text more manageable, I will simply present the median value for the calibrated age estimate, instead of repeating this probability estimate every time I mention an age, using the following format: 3185 cal BP. For those wishing to see the ^{14}C dates as originally reported, please consult the cited publications or refer to the online database, Ancient Maize Map (Blake et al. 2012).

22. Long et al. 1989. This date is given here as uncalibrated since the original phase dates are also uncalibrated. The calibrated phase dates for the Coxcatlán Phase would be approximately 7800–6150 cal BP, while the calibrated date for the earliest maize sample, from San Marcos Cave, is 5650–5050 cal BP. This means that the earliest maize is about 2,150 years younger than what was considered the maximum age for maize in the Tehuacán Valley—which is the age estimate that had usually been given as the starting date for the Coxcatlán Phase. The dates now show that, at a minimum, the estimate is 500 years younger than the youngest part of the phase.

23. Flannery 1986b:250.

24. Fritz 1994.

25. My colleague Bruce Benz and I have recently completed a maize AMS dating project in Latin America, funded by the Social Sciences and Humanities

Research Council of Canada (2007–2010). Working closely with archaeologists and paleoethnobotanists from the Latin America, Canada, and the United States, we dated ninety-two samples from fifteen different archaeological sites in six countries. We reported most of the new dates in our online website and database, the Ancient Maize Map (Blake et al. 2012).

26. Smith 2001, 2005; Piperno and Flannery 2001; Benz et al. 2006.

27. Dillehay et al. 2012; Spencer 2010.

28. Blake 2006.

29. Explore the Ancient Maize Map to see the locations of all of these sites and the range of samples from them: http://en.ancientmaize.com/locations/209. Huber and Van West 2005; Mabry 2008.

30. Blake et al. 1995; Clark 1994. This project was sponsored by the New World Archaeological Foundation of Brigham Young University and partially funded by the Social Sciences and Humanities Research Council of Canada. Work at these and several other sites in the Mazatán region continued through the mid-1990s, and the excavations are summarized in a volume edited by Richard Lesure (2011). More recently, Robert Rosenswig and his team recovered maize remains from the Early Formative site of Cuauhtémoc, just to the southeast of Mazatán, and the earliest of the samples they found dates to ca. 3430 cal BP (Rosenswig 2010), the same age as the earliest maize samples in the Mazatán region.

31. Scheffler 2008.

32. Blake et al. 2012. In work still underway, Hirth and Webster have since AMS-dated many more maize and other samples from El Gigante rockshelter and have been able to refine the stratigraphic contexts considerably.

33. Bradley and Vieja 1994.

34. Bonavia and Grobman 1989:839.

35. Unfortunately, both David Kelley and Duccio Bonavia recently passed away—but their substantial contributions live on in their publications, the collections they made, and the works of their students and colleagues, who continue their legacy.

36. Bonavia and Grobman 1989:838; Bonavia 1982, 2013; Kelley and Bonavia 1963.

37. Our project to sample the Los Gavilanes (Huarmey) maize in the Harvard collections was greatly facilitated by the logistical help of Steven LeBlanc and Jeffrey Quilter at the Harvard Peabody Museum of Archaeology and Ethnology. Ginny Popper, a close colleague and friend of mine, is an archaeologist and ethnobotanist at University of Massachusetts-Boston. She had previously analyzed the non-maize samples from Los Gavilanes (Popper 1982) and helped us with the sampling and lab work. All of our new AMS dates on samples from Los Gavilanes are reported online as part of the Ancient Maize Map (http://en.ancientmaize.com/sites/143). For more on the Los Gavilanes site see Bonavia 1982, 2013; Bonavia and Grobman 1989; Grobman and Bonavia 1978; Grobman et al. 1977; Kelley and Bonavia 1963.

38. Bradley 2015; Dull 2006; Horn 2006.

39. Pope et al. 2001; Pohl et al. 2007.

40. Teosinte pollen dating between 9500 and 6980 BP (ca. 10,800–7800 cal BP) has been recovered from Guilá Naquitz Cave. It is possible that this pollen

was from cultivated teosinte or very early maize. If this were the case, it would represent some of the earliest indirect evidence we have for any cultivated teosinte or maize pollen in Mexico (Piperno and Flannery 2001:2102).

41. Piperno et al. 1985; Dickau 2010; Dickau et al. 2007.

42. Piperno et al. 2000.

43. Perry et al. 2007.

44. Wilding 1967; Mulholland and Prior 1993. Dolores Piperno and her colleagues have used this AMS dating technique most frequently and successfully, and Piperno provides a comprehensive history of the method and a case example from Ecuador in the most recent edition of her text on phytolith analysis (2006:125–131).

45. Bray et al. 1987.

46. Piperno and Pearsall 1998; Stothert 1985.

47. Pearsall 1978, 2003; Pearsall and Piperno 1990.

48. Zarrillo 2012:213–214; Valdez 2008; 2013

49. Perry et al. 2006. This date is based on the conventional radiocarbon dating of a bulk sample of charcoal.

50. Mejía 2003.

51. Staller and Thompson 2002; Tykot and Staller 2002.

52. Piperno 2006; Pearsall and Piperno 1990; Staller 2003; Staller and Thompson 2002; Tykot and Staller 2002; Piperno 2003; Pearsall et al. 2004; Pearsall 2002.

53. Pearsall et al. 2004.

5. MAIZE THROUGH A MAGNIFYING GLASS

1. Deborah Pearsall's book *Paleoethnobotany: A Handbook of Procedures* (2000) is the best, most comprehensive guide to this field of study.

2. Pearsall 2000; Struever 1968; French 1971.

3. Minnis and LeBlanc 1976.

4. Minnis 1985, 1984.

5. Mangelsdorf 1974; Mangelsdorf et al. 1967; Mangelsdorf and Smith 1949.

6. Mangelsdorf 1974:113.

7. Mangelsdorf and Reeves 1939; Doebley 2001; Emshwiller 2006.

8. Beadle 1939, 1972; Doebley 2001.

9. Benz et al. 2006; Benz and Iltis 1990; Benz and Long 2000.

10. Flannery 1986a:8.

11. Benz 2001; Smith 2001; Piperno and Flannery 2001.

12. Purugganan and Fuller 2009.

13. Benz 2001:2104.

14. Iltis 2006:36, figure 3–10.

15. Benz 2001:2105.

16. Benz 2006; Iltis 2006; Smith 2001.

17. Beadle 1972.

18. As mentioned in chapter 4, direct AMS dating has placed the age of the earliest maize in the Tehuacán Valley at 5400 cal BP, rather than approximately

7,000 years old, as had been assumed by indirect dating (Mangelsdorf 1974:167).

19. Benz et al. 2006; Benz and Iltis 1990; Benz and Long 2000. Benz (pers. comm., April 2013) has now AMS dated about fifty more samples of maize cobs from the Tehuacán caves, bringing the total to over eighty-five. This goes a long way toward addressing some of the criticisms that have been leveled at Benz and Long's earlier study of the Coxcatlán maize collection, in which undated cobs were included. For example, Bruce Smith (2005) reanalyzed the Coxcatlán botanical materials, providing an enormous suite of AMS dates (twenty-eight on a range of plants, in addition to the six previous maize cob dates obtained by Long et al. in 1989), and showed that the cave deposits had undergone a great deal of disturbance, particularly those on the west side. He cautioned that it was almost impossible to have any confidence that plant remains found in a given occupation layer actually date to that layer. Smith's analysis of the radiocarbon dates showed that thirty-four of the seventy-one dated plant specimens (48 percent) migrated either upward or downward in the deposits—as a result of human pit-digging activities or rodent burrowing taking place long after the occupation layers had formed. Smith notes that while the general cultural sequence described by Johnson and MacNeish (1972) was largely correct, the dating by association of any individual artifact or plant fragment was likely to be unreliable. This is even more of a problem in the westernmost excavation units of the cave, where twenty-one of twenty-seven samples (78 percent) were significantly displaced. So for now, unless maize samples are directly dated using AMS or can be demonstrated to have come from unmixed, undisturbed deposits—in other words, where dating analyses have shown that directly dated specimens are in the correct stratigraphic context—we should be wary of morphological analyses that use samples that have been only dated by association.

20. Benz et al. 2006:77.

21. Wellhausen et al. 1952; Grobman et al. 1961. In *Corn: Its Origin, Evolution and Improvement*, Paul Mangelsdorf (1974:101–111) outlines the massive Races of Maize project, begun in 1943, to improve maize through selective plant breeding. An international team of botanists and agronomists began by collecting as many varieties of maize as they could find from all over Mexico and classifying them into thirty-two different races. This project eventually expanded to include all of the Americas, and Mangelsdorf (1974:105) estimated that the team defined approximately one hundred distinguishable races.

22. Benz et al. 2007.

23. Tuxill et al. 2010:469–470.

24. Tuxill et al. 2010:481.

6. MAIZE THROUGH A MICROSCOPE

1. Gamwell 2003; Breed and Conn 1936; Wainwright 2001.

2. Darwin 1846:5–6.

3. This expression was popularized in Rodgers and Hammerstein's musical *Oklahoma*, a 1943 Broadway hit. However, maize's claim to fame is being challenged by other species, such as elephant grass (*Miscanthus giganteus*),

which grows to almost 5 meters in height. Some of these grasses are being touted as alternatives to maize for the production of biofuels (Sanderson 2005).

4. Goldberg et al. 1993; McCormick 1993; Zhao et al. 2002.

5. Jacobs et al. 1999:593.

6. Katiyar and Sachan 1992. These values have been rounded.

7. Luna V. et al. 2001.

8. Piperno 2006:5.

9. Dolores Piperno's book *Phytoliths: A Comprehensive Guide for Archaeologists and Paleoecologists* (2006) provides the most comprehensive guide to understanding phytolith physiology, morphology, and methods of collection and analysis, and includes a summary of research applications around the world.

10. Some of the oldest grass phytoliths have been recently discovered in the fossilized droppings of dinosaurs called titanosaur sauropods, which lived in what is now India during the Late Cretaceous, approximately 70 million years ago (Prasad et al. 2005).

11. Pearsall and Piperno (1990) have done pioneering research on the differences between maize phytoliths and those of other species of tropical grasses. Recent work by Hart and Matson (2009) has shown that it is also possible to differentiate between maize phytoliths and those of various grasses that are found in the temperate northeastern United States.

12. Reichert 1913. Pearsall (2000) includes some of Reichert's superb photos in her paleoethnobotany handbook.

13. Horton et al. 2006:237–238.

14. Pearsall 2000:180.

15. Flannery 2002:250.

16. Pearsall 1988, 1995, 2003:214–215.

17. Bruce Smith (1994) has, in the past, argued that a more conservative approach to identifying the domestication of any plant species should rely on observations of traits of domestication, such as increasing seed size.

18. Aylor et al. 2005.

19. Aylor 2003.

20. Bragg 1969.

21. Neff et al. 2006a, 2006b.

22. Faegri et al. 1989; Piperno 2006.

23. Neff et al. 2006a.

24. Fearn and Liu 1995.

25. Eubanks 1997.

26. Ekdahl et al. 2004:figure 2.

27. Pearsall 1978, 2000.

28. Piperno 2006.

29. Loy et al. 1992.

30. Pearsall et al. 2004.

7. ELEMENTAL MAIZE

1. Many excellent charts available online show the periodic table of elements, but it is now possible to find superb interactive tables and charts that

illustrate the relationships among the elements and their isotopes (which are also called nuclides; the two terms are sometimes used interchangeably). I recommend the Interactive Chart of Nuclides, available at the website www.nndc .bnl.gov/chart/index.jsp, which is supported by the National Nuclear Data Center of the Brookhaven National Laboratory (NNDC-BNL 2015).

2. There are eight or nine even rarer isotopes of carbon. They each have six protons, but they vary in atomic weight, from eight neutrons at the light end to twenty-three neutrons at the heavy end. All of them are extremely unstable—lasting on average no more than a day and usually only a few seconds. These isotopes are thought to be the products of human-made nuclear reactions and are not believed to occur naturally.

3. The C3 cycle is also called the Benson-Calvin cycle, after Andrew Benson and Melvin Calvin, the two chemists who discovered and described this process in the late 1940s (Calvin and Benson 1949).

4. The C4 cycle is also known as the Hatch-Slack cycle, named after Marshall Hatch and Charles Slack, two Australian plant biochemists who, in the 1960s, were the first to comprehensively outline the differences between the photosynthesis pathways in plants that produce four carbon molecules (C4 cycle plants) and those that produce three carbon molecules (C3 cycle plants) (Slack and Hatch 1967). The C4 cycle is thought to be a more efficient way for plants to conserve water during **transpiration** in relatively hot, arid regions of the globe.

5. Keeley and Rundel 2003.

6. O'Leary 1988:329, figure 1.

7. Because of a process called fractionation, the stable isotope ratios of the consumers of these plants are depleted in ^{13}C, leading to predictably higher values, or offsets, compared to their source foods.

8. Vogel and van der Merwe 1977; van der Merwe and Vogel 1978.

9. Downward 2007; Allison 1952.

10. Michael Richards and Erik Trinkaus (2009) provide an excellent example of this change in trophic levels in the case of Neanderthals. Of course Neanderthals weren't consuming maize, or any C4 plants for that matter, but the same principle applies.

11. Pacbitun was excavated from 1984 to 1987 by Paul Healy and his team from Trent University, and the subsequent stable isotope analyses were carried out by Christine White and Henry Schwarcz, from Western University and McMaster University, respectively (White et al. 1993).

12. White 1997; White et al. 2006, 2001.

13. White et al. 2006:150.

14. White et al. 2006:150.

15. Pearsall 2003; McClung de Tapia 2000.

16. Diana Moreiras (2013) recently completed her Master's thesis, reanalyzing the samples from the Soconusco that Brian Chisholm and I had previously studied in the early 1990s (Blake et al. 1992b, 1992a). Her updated results provide the raw data for figure 7.3.

17. Chisholm and Blake 2006; Voorhies et al. 1991.

18. Moseley 1975, 1992, 2001.

19. Ericson et al. 1989.

20. Burger and van der Merwe 1990.

21. Tykot, Burger, and van der Merwe (2006) examined materials from one additional site, but no collagen was recovered from those samples, so we will just look at the patterns from the three mentioned in the text.

22. The $\delta^{13}C$ ‰ apatite values for a given sample are always much more positive than the collagen values (Tykot 2006).

23. Smalley and Blake 2003.

24. Chisholm et al. 1982; Ambrose and Norr 1993; Tykot 2006.

25. Katzenberg 2003.

26. Canal et al. 2008; Canal 2006.

8. GENETICALLY MODIFIED MAIZE THE OLD WAY

1. Doebley et al. 2006.

2. Jaenicke-Després et al. 2003.

3. In 2015 Rute da Fonseca, Bruce Smith, and their colleagues published a paper presenting new aDNA research on the movement of maize into the US Southwest, which began about 4000 cal BP. Analyzing the DNA from thirty-two archaeological samples—most of which come from sites mentioned in previous chapters (Tularosa Cave, Bat Cave, Tehuacán Caves) as well as from other sites in the United States that we have not discussed (including McEuen Cave and Turkey House Ruin), and one in northern Chile (Arica)—they found that the earliest wave of maize came into the US Southwest from Mexico along the highland route. Later, by about 2000 cal BP, new varieties of maize may have come into the United States via a Pacific coastal route.

4. Matsuoka et al. 2002.

5. Doebley et al. 1984; Doebley 2004:41.

6. There are four other species of teosinte: *Zea diploperennis, Zea luxurians, Zea nicaraguensis,* and *Zea perennis.* The species *Zea mays* includes the three subspecies mentioned in the text (*parviglumis, mexicana,* and *huehuetenangensis*) as well as domesticated *Zea mays* ssp. *mays.*

7. Matsuoka et al. 2002:6080.

8. They used a multivariate statistical analysis called principal component analysis (PCA).

9. Botanists call this sort of mixing "introgression."

10. Matsuoka et al. 2002.

11. Doebley 2004.

12. Dorweiler et al. 1993; Iltis 2006; Wang et al. 2005.

13. Doebley 2004.

14. Wang et al. (2005:718) state that: "For *tga1,* we have identified a set of seven SNPs, one or some combination of which represent the causative site(s). On the basis of the available evidence, our preferred hypothesis is that the K→N substitution controls the phenotypic difference between maize and teosinte."

15. Wang et al. 2005:714, 718.

16. Boerjan et al. 2003; Wang et al. 2005.

17. Doebley 2004.

18. Iltis 1983, 2000.

19. Iltis 2000:17, 2006; Doebley 2004:49–51.

20. Iltis 2006.

21. Doebley et al. 2006.

22. Doebley et al. 2006.

23. Doebley et al. (2006: 1311, figure 2) caution, however, "that an exceptionally strong domestication bottleneck could leave little variation in neutral genes. In that case, it may be very difficult to distinguish selected from neutral loci."

24. Wright and Gaut 2005.

25. Crichton 1990.

26. Ancient-DNA specialists think it unlikely that DNA can preserve in a decipherable form beyond several tens of thousands of years because the chemical deterioration process is inevitable. However, their decay may be slowed down through methods such as perpetual freezing (Pääbo et al. 2004).

27. Green et al. 2006. It seems that every month or two, new ancient-DNA studies are announced in the global media, and often they shed considerable new light on the genetic makeup, migrations, and adaptations of our human ancestors.

28. Goloubinoff et al. 1993.

29. Jaenicke-Després and Smith 2006:88–89; Wang et al. 1998.

30. Pääbo et al. 2004.

31. Jaenicke-Després and Smith 2006.

32. Doebley et al. 2006. Rute da Fonseca and her team have updated the DNA analysis of ancient US Southwest and Mexican maize samples; see note 3 (da Fonseca et al. 2015).

33. Jaenicke-Després and Smith 2006.

34. Matsuoka et al. 2002.

35. Piperno and Flannery 2001.

36. Wang et al. 2005.

37. Przeworski 2003.

38. Edward Buckler, Jeffrey Ross-Ibarra, John Doebley, and Doreen Ware, among many others, all working out of their respective university labs and partnered with the US Department of Agriculture, are the principal researchers guiding the HapMap project.

39. Panzea 2014.

40. Gore et al. 2009.

41. Chia et al. 2012; Hufford et al. 2012.

9. DAILY TOOLS AND SACRED SYMBOLS

1. Biskowski and Watson 2013.

2. Carter 1980:37.

3. Bruhns 1994:91; Evans 2013:72–74.

4. Duffy 2011.

5. Blake 2010:178–181.

6. Clark and Cheetham 2002; Clark et al. 2007.

7. Clark 1994:236.

8. There are many excellent works that discuss the manufacturing techniques and properties of preindustrial pottery from a global archaeological perspective. Foremost among these are *Pottery Analysis: A Sourcebook* by Prudence Rice (2005) and *Approaches to Archaeological Ceramics* by Carla Sinopoli (1991).

9. Karttunen 1992. Also see the online Nahuatl dictionary, http://whp .uoregon.edu/dictionaries/nahuatl/index.lasso.

10. Katz et al. 1974:766.

11. Bressani et al. 1958; Bressani and Scrimshaw 1958. Nixtamalization is the process of cooking maize in a solution consisting of water and any of a group of alkali compounds made using mineral lime, lye, burned marine shells, or wood or other plant ash. When added to water, these substances make an alkaline solution of calcium hydroxide, potassium hydroxide, or sodium hydroxide (Beck 2001:189).

12. Brenton and Paine 2000.

13. Katz et al. 1974.

14. Beck 2001; Snow 1990.

15. It is possible that among the peoples of the southwestern United States, stone griddles were used for making a type of maize bread similar to Hopi *piki,* which is made with a thin maize batter that is spread across a hot stone griddle and briefly cooked, much more like a crepe than a tortilla (Karl Taube, pers. comm., August 28, 2014; Moerman 1998).

16. Clark and Blake 1994; Clark 1994.

17. Clark et al. 2007.

18. These Cunil Horizon colanders are quite small compared to the larger *olla*-shaped colanders that people used through the Classic and Postclassic periods and up to the present day. David Cheetham (2010a:358–359) shows that the Cunil Horizon colanders were bowl-shaped and, on average, were about 24.5 centimeters in diameter and held a volume of 3.3 liters. In contrast, the Cunil boiling pots (*ollas* and *tecomates*) ranged from an estimated 21 to 63 liters in volume. Therefore, if the small colanders were used to drain and rinse the contents of the cooking pots, they would likely have overflowed. Cheetham notes that the Cunil colanders are often red-slipped on their outer rims. This is a pattern I also noticed on small, finer colanders from Postclassic Canajasté, as well as colanders found in the present-day Maya communities of Chanal and Aguacatenango in highland Chiapas. I suspect that fine, decorated colanders might have been used to strain *atole*—a ceremonial maize and chocolate drink— rather than to drain and rinse nixtamalized maize for grinding into dough. However, David Cheetham (pers. comm., June 27, 2014) thinks this hypothesis unlikely since the hole size of the colanders match that of boiled maize strainers rather than ceremonial *atole* sieves.

19. Fournier 1998.

20. Fish 2004:129; Beck 2001.

21. Myers 2006; Hart 2008, 2014.

22. Boyd and Surrette 2010; Largy and Morenan 2008; Spiess and Cranmer 2001.

23. Myers 2006:515–516.

24. Andrews 2000.

25. Coe 1962. The main goal of Coe's brief 1962 piece in *American Antiquity* was to suggest that a very early ceramic bottle from the Kotosh site in highland Peru dating to the late Initial period (about 900–1000 BCE) had a depiction of maize that was very similar to and perhaps inspired by Olmec iconography. The reasonable criticisms of this hypothesis (Lanning 1963) don't diminish the contribution of Coe's insight that this particular maize symbol, often shown as emanating from the cleft in the head of an Olmec god, could be interpreted as the representation of the Maize God: "Taking this as a lead, one might suggest that the cranial cleft of these snarling monstrosities may have been the mark of the sprouting maize, so intimately connected with the coming of the rains. Thus, the were-jaguar may have been lord of the maize as well" (Coe 1962:580).

26. Taube 1995:39–41; Coe 1968; Joralemon 1971.

27. Killion 2013.

28. Taube 2004.

29. Rodríguez Martínez et al. 2006:1613.

30. Cheetham 2010b.

31. Taube 1995:75–76. Teopantecuanitlán was discovered and excavated by Guadalupe Martínez Donjuán (1985, 1994) in the 1980s, and it is one of the best preserved early Olmec influenced sites in western Mexico.

32. Taube 1995:41–44, 2004:30, 107.

33. Taube 1995:70, figure 23.

34. Saturno et al. 2006.

35. Saturno et al. 2005; Taube 2010.

36. Eubanks 1999.

37. Berezkin 1980; Eubanks 1999:30–31; Bawden 2004:123. The introduction and spread of maize throughout much of Africa, which began in the late 1500s, provides interesting contrasts and parallels to the South American example. As historian James McCann (2005) documents, maize was most frequently adopted in Africa simply as a new food crop that could supplement traditional sources of carbohydrates. Unlike Mesoamericans and, to a lesser extent, South American peoples, Africans rarely regarded maize as sacred, and, with few exceptions, the plant was never represented as a deity or linked to origin stories.

38. Benson 1997:126, 2012; Burger 1997. Mary Eubanks (1999, 2001a) argues that some of these three-dimensional Moche portrayals of maize were made by pressing an actual ear of maize into the clay that was used to make the molds from which the vessels were cast.

39. Moseley 2001:8.

40. Moseley et al. 2005.

41. Taube 2000. See Hays-Gilpin and Hegmon (2005) for a convincing analysis of the variation among Puebloan peoples' use of maize symbolism. The two argue, for example, that the Mimbres people seldom depicted maize on their pottery but were much more concerned with showing themes related to hunting and other activities: "Although corn agriculture was key to subsistence in the Mimbres region, and some Hopi clans may have distant ancestors from this

area, we have virtually no evidence that corn was depicted in Mimbres pottery, rock art, or other media, even as an abstract visual metaphor. In the Mimbres tradition, and among many other Pueblo ancestors, corn may have been 'taken for granted,' not because it was unimportant, but because it was pervasive" (Hays-Gilpin and Hegmon 2005:107).

42. Taube 2000:324.

43. Taube 2000:326.

44. Karl Taube cites Polly Schaafsma, who recognized the link between the undulating maize stalk and the symbolic representation of lightning emanating from the cloud. And, because the maize ears appear flat-topped and have no silk protruding from them, they more closely resemble celts—an ancient symbol for the ear of maize—than realistic portrayals of maize.

45. See Bruce Smith's *Rivers of Change* (1992) for a comprehensive discussion of the long history of agriculture in eastern North America.

46. Knight 1986:679.

47. Hudson 1984:19–20.

48. Recinos et al. 1950. See Evon Z. Vogt's books *Zinacantan* (1969) and *Tortillas for the Gods* (1976) for some of the most comprehensive discussions and descriptions of maize's fundamental social, ritual, and symbolic importance to the contemporary Tzotzil Maya people of Zinacantán in highland Chiapas. In fact, it was the latter volume that provided the inspiration for the present book's title.

49. Geoffrey McCafferty, pers. comm., February 10, 2012.

50. Sheets et al. 2011, 2012.

Glossary

The following definitions are, with few exceptions, direct quotations from NHGRI 2014; Horton et al. 2006; UNC-Herbarium-Plant Information Center 2015; GHR 2007; Doebley 2003; Mauseth 2008; vPlants 2009; and Scitable 2015. Most of these references are available online, so readers can easily access these or similar websites to obtain more detailed descriptions or discussions as necessary. Wikipedia is also a good source for definitions of some of the terms, both archaeological and botanical, that appear in the text but are not included in this glossary.

ABSCISSION LAYER. A layer of cells connecting adjacent plant parts, such as seeds along a rachis or leaves on a branch. When mature, the abscission cells become brittle and break apart, providing a natural dispersal mechanism for the seeds or leaves. In domesticated cereals the abscission layer remains tough and does not break apart when the seeds are mature, thereby allowing farmers to harvest the grains more easily.

ALLELE. Variants of the same gene with small differences in the sequence of DNA bases.

AMS (ACCELERATOR MASS SPECTROMETRY) DATING. A radiocarbon dating method that combines a particle accelerator and a mass spectrometer to directly measure the relative proportions of ^{12}C and ^{14}C atoms in a sample. The process produces accurate measurements using very small quantities of material—unlike conventional radiocarbon dating, which measures the amount of beta decay particles and requires much larger samples.

ANNUAL PLANT. Plants that live and reproduce for only one year, in contrast to perennial plants, which may live and reproduce for many years.

APATITE. The phosphate mineral portion of bone and teeth, which, together with collagen, the protein portion of bone and teeth, make up the bulk of these organic structures.

APICAL. Appearing at the top, tip, or end of a plant structure.

BASE PAIR (BP) SEGMENT. Two bases which together form a "rung" of the DNA "ladder." A DNA nucleotide is made of a molecule of sugar, a molecule of phosphoric acid, and a molecule called a base. The bases are the "letters" that spell out the genetic code. In DNA, the code letters are A (adenine), T (thymine), G (guanine), and C (cytosine). In base pairing, adenine always pairs with thymine, and guanine always pairs with cytosine (except in RNA, where A pairs with U (uracil).

BRACT. A modified, usually reduced, leaf in the inflorescence (flower) of a plant.

CAL BP (CALIBRATED YEARS BEFORE PRESENT). Cal BP refers to calendar years before present, which, by convention, is taken to be 1950. Calibration of radiocarbon dates is a means of adjusting the age of a radiocarbon measurement by matching it to the corresponding graph of known tree-ring dates, yielding a true calendrical age. Because radiocarbon production rates in the atmosphere have varied over the course of centuries and millennia, the ^{14}C age of a sample can be much younger or older than its calendrical age.

COLLAGEN. An abundant protein that makes up a large part of the organic tissues of many animals, including all mammals. Collagen can be extracted from bone, teeth, hair, and other tissue for the purpose of stable isotope analysis.

CUPULE. Fused, hardened involucral bracts that hold a plant's flowers—somewhat resembling a cup.

DEMIC DIFFUSION. Expansion of a farming population into new regions through population growth.

DEOXYRIBONUCLEIC ACID (DNA). Polymers, or chains of molecules, inside cells that carry genetic information and pass it from one generation to the next.

DISTICHOUS. Opposite or alternate arrangement of leaves or other plant structures along an axis, such as a stem. Also called "two-ranked."

EXINE. The tough outer layer of pollen grains. Exine shapes and surface textures are distinct among most species of plants, allowing those who study pollen—palynologists—to determine the presence of specific plants in ancient deposits.

GENE. A segment of DNA that is transcribed or that encodes a protein, for example, an enzyme.

GENOME. A complete set of the genetic information contained within an organism.

GENOTYPE. The genetic characteristics of a trait or set of traits in a given organism. (Compare to PHENOTYPE.)

GENOTYPING. Characterizing the specific set of alleles at different DNA loci.

GLUME. A bract, usually occurring in pairs, at the base of a grass spikelet.

HAPLOTYPE. An organism's inherited cluster of SNPs—single nucleotide polymorphisms.

INDEL. An insertion or deletion of a section of genetic code—sets of base pairs—allowing researchers to identify specific mutations along a strand of DNA.

INFLORESCENCE. The cluster of flowers attached to the stem of a plant. In maize and teosinte, the male, or staminate, inflorescence, called the tassel, grows at the tip of the stalk and produces pollen. The female, or pistillate, inflorescence grows farther down on one of the stalk's nodes and produces the seeds.

INTROGRESSION. In plant genetics, the process of interbreeding whereby genetic material from a closely related wild species is introduced, either purposefully or accidentally, into the gene pool of a domesticated species.

INVOLUCRE. A group or cluster of bracts subtending (supporting) an inflorescence. In maize, these form the cupule.

ISOZYME. One of various structurally related forms of an enzyme, each having the same mechanism but with differing chemical, physical, or immunological characteristics.

LANDRACE. A local or regional population of maize (or any other plant species) that exhibits greater internal similarity than similarity to other maize populations, whether nearby or distant.

LOCUS. The physical site or location of a specific gene on a chromosome.

MICROSATELLITES. Repeating segments of DNA that are usually two to five nucleotides in length. Also called "simple sequence repeats" (SSRs) or "simple tandem repeats" (STRs), microsatellites are usually found in the noncoding regions of the genome. They are ideal genetic markers, because the high degree of variability in repeat number can be used to determine the level of genetic relatedness between individuals within a species (whether humans or maize).

MIDDEN. An archaeological deposit of food remains and general refuse accumulated around a dwelling or residential area. Shell heaps are often referred to as middens.

NUCLEOTIDE. One of the structural components, or building blocks, of DNA and RNA. A nucleotide consists of a base (in DNA, one of four chemicals—adenine (A), thymine (T), guanine (G), and cytosine (C)—while in RNA, uracil (U) is found instead of thyamine) plus a molecule of sugar and a molecule of phosphoric acid.

PERENNIAL PLANT. Plants that live and reproduce for many years, in contrast to annual plants, which typically live and reproduce for only one year.

PHENOTYPE. The physical appearance of an organism based on its genotype—or genetic makeup.

PHYTOLITH. Small, distinctively shaped silica bodies that form both within and between the cell structures of many plants. In some cases, they help give plant structures their rigidity and perhaps also provide protection against some pests and browsers. The distinctive shapes of these durable particles can be found in ancient archaeological sites and allow for the identifications of plant species present in ancient times—even in the absence of any other type of plant remains.

POLYMERASE CHAIN REACTION (PCR). A method of amplifying the amount of DNA in a sample. A small sequence of DNA, for example, a gene, can be isolated and copied, allowing for more detailed study of its properties.

POLYSTICHOUS. Leaves, flowers or other plant structures that are arranged in two or more rows.

PRIMER. A molecule (such as a short strand of RNA or DNA) whose presence is required for the formation of another molecule (such as a longer chain of

DNA). Primers are synthetically constructed, using polymerase chain reaction (PCR) technology, by molecular biologists to replicate large enough quantities of an individual DNA sample to identify and study it.

PRIMORDIUM (PL. PRIMORDIA). The cells in a plant that begin the first stage of development and growth of an organ, such as the tip of a branch, leaf, or flower.

RACHIS. The central axis of a compound stem, along which the leaves, flowers, or seed heads are attached.

RIBONUCLEIC ACID (RNA). RNA transmits genetic information from DNA to proteins produced by the cell.

SNP. Pronounced "snip," this abbreviation stands for "single nucleotide polymorphism." A SNP is a nucleotide that can vary between individuals of the same species. SNP variants that occur in coding regions of a gene are considered to be alleles of that gene.

SPIKE. An unbranched inflorescence in which the flowers are sessile—that is, without stems—and grow on an elongated axis.

SPIKELET. A small spike branching off the main spike, which, in grasses, forms the inflorescence or florets, which are in turn supported by bracts. In maize, the cupule contains two spikelets, or florets, which become the paired kernels. In teosinte, the cupule contains a single spikelet.

SPOROPOLLENIN. A tough polymer that constitutes the exine (outer) layer of pollen grains and some spores.

SYMPATRIC. Different species or subspecies that inhabit the same geographic region or zone.

TRANSPIRATION. The process whereby water, which is normally taken up by the roots of a plant, is given off as water vapor by the pores in its leaves.

References

Allison, Samuel K. 1952. *Arthur Jeffrey Dempster, 1886–1950: A Biographical Memoir*. Washington, DC: National Academy of Sciences.

Ambrose, Stanley H., and Lynette Norr. 1993. "Experimental Evidence for the Relationship of the Carbon Isotope Ratios of Whole Diet and Dietary Protein to Those of Bone Collagen and Carbonate." In *Prehistoric Human Bone: Archaeology at the Molecular Level,* edited by J.B. Lambert and G. Grupe, pp. 1–37. Berlin: Springer-Verlag.

Ammerman, A.J., and L.L. Cavalli-Sforza. 1984. *The Neolithic Transition and the Genetics of Populations in Europe*. Princeton, NJ: Princeton University Press.

Andrews, Tamra. 2000. *Nectar and Ambrosia: An Encyclopedia of Food in World Mythology*. Santa Barbara, CA: ABC-CLIO.

Arnold, James R., and Willard F. Libby. 1951. "Radiocarbon Dates." *Science* 113 (2927):111–120.

Aylor, Donald E. 2003. "Rate of Dehydration of Corn (*Zea mays* L.) Pollen in the Air." *Journal of Experimental Botany* 54 (391):2307–2312.

Aylor, Donald E., Baltazar M. Baltazar, and John B. Schoper. 2005. "Some Physical Properties of Teosinte (*Zea mays* subsp. *parviglumis*) Pollen." *Journal of Experimental Botany* 56 (419):2401–2407.

Bawden, Garth. 2004. "The Art of Moche Politics." In *Andean Archaeology,* edited by H. Silverman, pp. 116–129. Malden, MA: Blackwell Publishing.

Beadle, George W. 1939. "Teosinte and the Origin of Maize." *Journal of Heredity* 30:245–247.

———. 1972. "The Mystery of Maize." *Field Museum of Natural History, Bulletin* 43 (10):2–11.

———. 1980. "The Ancestry of Corn." *Scientific American* 242 (1):112–119.

Beck, Margaret. 2001. "Archaeological Signatures of Corn Preparation in the US Southwest." *The Kiva* 67 (2):187–218.

Bellwood, Peter, and Colin Renfrew, eds. 2003. *Examining the Farming/Language Dispersal Hypothesis*. Cambridge, UK: McDonald Institute for Archaeological Research, University of Cambridge.

Bellwood, Peter S. 2005. *The First Farmers: The Origins of Agricultural Societies*. London: Blackwell Publishing.

Bennett, C. L., R. P. Beukens, M. R. Clover, H. E. Gove, R. B. Liebert, A. E. Litherland, et al. 1977. "Radiocarbon Dating Using Electrostatic Accelerators: Negative Ions Provide the Key." *Science* 198 (4316):508–510.

Bennetzen, J. L., E. Buckler, V. Chandler, J. Doebley, J. Dorweiler, B. Gaut, et al. 2001. "Genetic Evidence and the Origin of Maize." *Latin American Antiquity* 12 (1):84–86.

Benson, Elizabeth P. 1997. "Moche Art: Myth, History, and Rite." In *The Spirit of Ancient Peru: Treasures from the Museo Arqueológico Rafael Larco Herrera*, edited by K. Berrin, pp. 41–49. New York: Thames and Hudson.

———. 2012. *The Worlds of the Moche on the North Coast of Peru*. Austin: University of Texas Press.

Benz, Bruce, Hugo Perales, and Stephen Brush. 2007. "Tzeltal and Tzotzil Farmer Knowledge and Maize Diversity in Chiapas, Mexico." *Current Anthropology* 48 (2):289–300.

Benz, Bruce F. 2000. "The Origins of Mesoamerican Agriculture: Reconnaissance and Testing in the Sayula-Zacoalco Lake Basin." FAMSI Online Reports. www.famsi.org/reports/99074/99074Benz01.pdf , accessed March 4, 2015.

———. 2001. "Archaeological Evidence of Teosinte Domestication from Guilá Naquitz, Oaxaca." *Proceedings of the National Academy of Sciences, USA* 98 (4):2104–2106.

———. 2006. "Maize in the Americas." In *Histories of Maize: Multidisciplinary Approaches to the Prehistory, Biogeography, Domestication, and Evolution of Maize,* edited by J. E. Staller, R. H. Tykot, and B. F. Benz, pp. 9–20. Amsterdam: Academic Press.

Benz, Bruce F., Li Cheng, Steven W. Leavitt, and Chris Eastoe. 2006. "El Riego and Early Maize Agricultural Evolution." In *Histories of Maize: Multidisciplinary Approaches to the Prehistory, Biogeography, Domestication, and Evolution of Maize,* edited by J. E. Staller, R. H. Tykot, and B. F. Benz, pp. 73–82. Amsterdam: Academic Press.

Benz, Bruce F., and Hugh H. Iltis. 1990. "Studies in Archaeological Maize I: The 'Wild' Maize from San Marcos Cave Reexamined." *American Antiquity* 55 (3):500–511.

Benz, Bruce F., and Austin Long. 2000. "Prehistoric Maize Evolution in the Tehuacan Valley." *Current Anthropology* 41 (3):459–465.

Berezkin, Yuri E. 1980. "An Identification of Anthropomorphic Mythological Personages in Moche Representations." *Ñawpa Pacha: Journal of Andean Archaeology* 18:1–26.

Biskowski, Martin, and Karen D. Watson. 2013. "Changing Approaches to Maize Preparation at Cerro Portezuelo." *Ancient Mesoamerica* 24 (1):213–223.

Blake, Michael. 2006. "Dating the Initial Spread of *Zea mays*." In *Histories of Maize: Multidisciplinary Approaches to the Prehistory, Biogeography,*

Domestication, and Evolution of Maize, edited by J. E. Staller, R. H. Tykot, and B. F. Benz, pp. 55–71. Amsterdam: Academic Press.

———. 2010. *Colonization, Warfare, and Exchange at the Postclassic Maya Site of Canajasté, Chiapas, Mexico.* Papers of the New World Archaeological Foundation, no. 70. Provo, UT: Brigham Young University.

Blake, Michael, Bruce Benz, Nicholas Jakobsen, Ryan Wallace, Sue Formosa, Kisha Supernant, et al. 2012. Ancient Maize Map, Version 1.1: An Online Database and Mapping Program for Studying the Archaeology of Maize in the Americas. http://en.ancientmaize.com/. Vancouver: Laboratory of Archaeology, University of British Columbia.

Blake, Michael, Brian S. Chisholm, John E. Clark, Barbara Voorhies, and Michael W. Love. 1992a. "Prehistoric Subsistence in the Soconusco Region." *Current Anthropology* 33 (1):83–94.

Blake, Michael, Brian S. Chisholm, John. E. Clark, and Karen Mudar. 1992b. "Non-Agricultural Staples and Agricultural Supplements: Early Formative Subsistence in the Soconusco Region, Mexico." In *Transitions to Agriculture in Prehistory,* edited by A. B. Gebauer and T. D. Price, pp. 133–151. Madison: Prehistory Press.

Blake, Michael, John E. Clark, Barbara Voorhies, George Michaels, Michael W. Love, Mary E. Pye, et al. 1995. "Radiocarbon Chronology for the Late Archaic and Formative Periods on the Pacific Coast of Southeastern Mesoamerica." *Ancient Mesoamerica* 6:161–183.

Boerjan, Wout, John Ralph, and Marie Baucher. 2003. "Lignin Biosynthesis." *Annual Review of Plant Biology* 54 (1):519–546.

Bommert, Peter, Namiko Satoh-Nagasawa, and David Jackson. 2013. "Quantitative Variation in Maize Kernel Row Number is Controlled by the *FASCIATED EAR2* Locus." *Nature Genetics* 45 (3):334–337.

Bonavia, Duccio. 1982. *Precerámico Peruano: Los Gavilanes: Mar, desierto y oásis en la historia del hombre.* Lima: Corporación Financiera de Desarrollo S.A. and Instituto Arqueológico Alemán.

———. 2013. *Maize: Origin, Domestication, and Its Role in the Development of Culture.* Cambridge, UK: Cambridge University Press.

Bonavia, Duccio, and Alexander Grobman. 1989. "Preceramic Maize in the Central Andes: A Necessary Clarification." *American Antiquity* 54 (4):836–840.

Bourdieu, Pierre. 1984. *Distinction: A Social Critique of the Judgement of Taste.* Translated by Richard Nice. Cambridge, MA: Harvard University Press.

Bowman, Sheridan. 1990. *Interpreting the Past: Radiocarbon Dating.* Berkeley: University of California Press.

Boyd, Matthew, and Clarence Surrette. 2010. "Northernmost Precontact Maize in North America." *American Antiquity* 75 (1):117–133.

Bradley, J. E., and T. Vieja. 1994. "An Archaic and Early Formative Site in the Arenal Region, Costa Rica." In *Archaeology, Volcanism, and Remote Sensing in the Arenal Region, Costa Rica,* edited by P. D. Sheets and B. R. McKee, pp. 73–86. Austin: University of Texas Press.

Bradley, Raymond S. 2015. *Paleoclimatology: Reconstructing Climates of the Quaternary.* 3rd ed. San Diego: Academic Press.

Bragg, Louis H. 1969. "Pollen Size Variation in Selected Grass Taxa." *Ecology* 50 (1):124–127.

Bray, Warwick, L. Herrera, M.C. Schrimpff, P. Botero, and J.G. Monsalve. 1987. "The Ancient Agricultural Landscape of Calima, Colombia." In *Pre-Hispanic Agricultural Fields in the Andean Region*, edited by W.M. Denevan, K. Mathewson, and G. Knapp, pp. 443–481. Oxford: BAR International Series.

Breed, Robert S., and H.J. Conn. 1936. "The Status of the Generic Term *Bacterium* Ehrenberg 1828." *Journal of Bacteriology* 31 (5):517–518.

Brenton, Barrett P., and Robert R. Paine. 2000. "Pellagra and Paleonutrition: Assessing the Diet and Health of Maize Horticulturists through Skeletal Biology." *Nutritional Anthropology* 23 (1):2–9.

Bressani, R., R. Paz y Paz, and N.S. Scrimshaw. 1958. "Chemical Changes in Corn During Preparation of Tortillas." *Journal of Agricultural and Food Chemistry* 6:770–774.

Bressani, R., and N.S. Scrimshaw. 1958. "Effect of Lime Treatment on in Vitro Availability of Essential Amino Acids and Solubility of Protein Fractions in Corn." *Journal of Agricultural and Food Chemistry* 6:774–778.

Bronk Ramsey, C. 2008. "Radiocarbon Dating: Revolutions in Understanding." *Archaeometry* 50 (2):249–275.

Bronk Ramsey, Christopher. 2014. Oxcal 4.2. https://c14.arch.ox.ac.uk/Oxcal/Oxcal.html. Oxford: Oxford Radiocarbon Accelerator Unit.

Bruhns, Karen Olsen. 1994. *Ancient South America*. Cambridge, UK: Cambridge University Press.

Bruman, Henry J. 2000. *Alcohol in Ancient Mexico*. Salt Lake City: University of Utah Press.

Burger, Richard L. 1997. "Life and Afterlife in Pre-Hispanic Peru." In *In the Spirit of Ancient Peru: Treasures from the Museo Arqueológico Rafael Larco Herrera*, edited by K. Berrin, pp. 21–32. New York: Thames and Hudson.

Burger, Richard L., and Nikolaas J. van der Merwe. 1990. "Maize and the Origin of Highland Chavín Civilization: An Isotopic Perspective." *American Anthropologist* 92 (1):84–95.

Byers, Douglas S., ed. 1967. *The Prehistory of the Tehuacán Valley*. Vol. 1: *Environment and Subsistence*. Austin: University of Texas Press.

Calvin, Melvin, and Andrew A. Benson. 1949. "The Path of Carbon in Photosynthesis IV: The Identity and Sequence of the Intermediates in Sucrose Synthesis." *Science* 109 (2824):140–142.

Canal, Cecilia Maria. 2006. "Stable Carbon Isotope Analysis and Maize-Stalk Beer Diet in Rats: Implications for the Origins of Maize." Master's thesis, University of British Columbia.

Canal, Cecilia Maria, Brian S. Chisholm, and Michael Blake. 2008. "Maize, Beer, and Rats: Is Maize Stalk Beer Consumption Isotopically Visible in Bones?" Poster presented at the 73rd Annual Meeting of the Society for American Archaeology. Vancouver, BC.

Carter, George F. 1980. "The Metate: An Early Grain-Grinding Implement in the New World." In *Early Native Americans: Prehistoric Demography, Economy, and Technology*, edited by D.L. Browman, pp. 21–39. The Hague: Mouton Publishers.

Ceja Tenorio, Jorge F. 1985. *Paso de La Amada: An Early Preclassic Site in the Soconusco, Chiapas.* Papers of the New World Archaeological Foundation, no. 49. Provo, UT: Brigham Young University.

Chantarudee, Atip, Preecha Phuwapraisirisan, Kiyoshi Kimura, Masayuki Okuyama, Haruhide Mori, Atsuo Kimura, et al. 2012. "Chemical Constituents and Free Radical Scavenging Activity of Corn Pollen Collected from *Apis Mellifera* Hives Compared to Floral Corn Pollen at Nan, Thailand." *BMC Complementary and Alternative Medicine* 12:45.

Cheetham, David. 2010a. "Corn, Colanders, and Cooking: Early Maize Processing in the Maya Lowlands and Its Implications." In *Pre-Columbian Foodways: Interdiscipilanary Approaches to Food, Culture, and Markets in Ancient Mesoamerica,* edited by J. E. Staller and M. Carrasco, pp. 345–368. New York: Springer.

———. 2010b. "Cultural Imperatives in Clay: Early Olmec Carved Pottery from San Lorenzo and Cantón Corralito." *Ancient Mesoamerica* 21 (1):165–185.

Chia, Jer-Ming, Chi Song, Peter J. Bradbury, Denise Costich, Natalia de Leon, John Doebley, et al. 2012. "Maize HapMap2 Identifies Extant Variation from a Genome in Flux." *Nature Genetics* 44 (7):803–807.

Chisholm, Brian, and Michael Blake. 2006. "Isotope Analysis and Subsistence in the Soconusco Region, Mexico." In *Histories of Maize: Multidisciplinary Approaches to the Prehistory, Biogeography, Domestication, and Evolution of Maize,* edited by J. E. Staller, R. H. Tykot, and B. F. Benz, pp. 162–172. Amsterdam: Academic Press.

Chisholm, Brian S., and R.G. Matson. 1991. "Basketmaker II Subsistence: Carbon Isotopes and Other Dietary Indicators from Cedar Mesa, Utah." *American Antiquity* 56 (3):444–459.

Chisholm, Brian S., D. Erle Nelson, and Henry P. Schwarcz. 1982. "Stable-Carbon Isotope Ratios as a Measure of Marine Versus Terrestrial Protein in Ancient Diets." *Science* 216 (4550):1131–1132.

Clark, John E. 1994. "The Development of Early Formative Rank Societies in the Soconusco, Chiapas, Mexico." PhD diss., University of Michigan.

Clark, John E., and Michael Blake. 1994. "The Power of Prestige: Competitive Generosity and the Emergence of Rank Societies in Lowland Mesoamerica." In *Factional Competition and Political Development in the New World,* edited by E. M. Brumfiel and J. W. Fox, pp. 17–30. Cambridge, UK: Cambridge University Press.

Clark, John E., and David Cheetham. 2002. "Mesoamerica's Tribal Foundations." In *The Archaeology of Tribal Societies,* edited by W. A. Parkinson, pp. 278–339. Ann Arbor: International Monographs in Prehistory.

Clark, John E., Mary E. Pye, and Dennis Gosser. 2007. "Thermolithics and Corn Dependency in Mesoamerica." In *Archaeology, Art, and Ethnogenesis in Mesoamerican Prehistory: Papers in Honor of Gareth W. Lowe,* edited by L. S. Lowe and M. E. Pye, pp. 15–42. Papers of the New World Archaeological Foundation, no. 68. Provo, UT: Brigham Young University.

Coe, Michael D. 1962. "An Olmec Design on an Early Peruvian Vessel." *American Antiquity* 27 (4):579–580.

———. 1968. *America's First Civilization*. New York: American Heritage.

Coltrain, Joan Brenner, Joel C. Janetski, and Shawn W. Carlyle. 2007. "The Stable- and Radio-Isotope Chemistry of Western Basketmaker Burials: Implications for Early Puebloan Diets and Origins." *American Antiquity* 72 (2):301–321.

Crichton, Michael. 1990. *Jurassic Park*. New York: Alfred A. Knopf.

da Fonseca, Rute R., Bruce D. Smith, Nathan Wales, Enrico Cappellini, Pontus Skoglund, Matteo Fumagalli, et al. 2015. "The Origin and Evolution of Maize in the Southwestern United States." *Nature Plants,* 1 (1):1–5.

Darwin, Charles. 1846. *Journal of Researches into the Natural History and Geology of the Countries Visited during the Voyage of H.M.S. Beagle, round the World under the Command of Capt. Fitz Roy, R.N.* New York: Harper and Brothers.

Díaz del Castillo, Bernal. 1912. *The True History of the Conquest of New Spain*. Translated by A. P. Maudslay. London: Hakluyt Society.

———. 1963. *The Conquest of New Spain*. Translated by J. M. Cohen. Harmondsworth: Penguin Books.

Dickau, Ruth. 2010. "Microbotanical and Macrobotanical Evidence of Plant Use and the Transition to Agriculture in Panama." In *Integrating Zooarchaeology and Paleoethnobotany: A Consideration of Issues, Methods, and Cases,* edited by A. M. VanDerwarker and T. M. Peres, pp. 99–134. New York: Springer.

Dickau, Ruth, Anthony J. Ranere, and Richard G. Cooke. 2007. "From the Cover: Starch Grain Evidence for the Preceramic Dispersals of Maize and Root Crops into Tropical Dry and Humid Forests of Panama." *Proceedings of the National Academy of Sciences, USA* 104 (9):3651–3656.

Diehl, Michael W. 2005. "Morphological Observations on Recently Recovered Early Agricultural Period Maize Cob Fragments from Southern Arizona." *American Antiquity* 70 (2):361–375.

Dillehay, Tom D., Duccio Bonavia, Steven Goodbred, Mario Pino, Victor Vasquez, Teresa Rosales Tham, et al. 2012. "Chronology, Mound-Building and Environment at Huaca Prieta, Coastal Peru, from 13 700 to 4000 Years Ago." *Antiquity* 86 (2012):48–70.

Doebley, John. 1990. "Molecular Evidence and the Evolution of Maize." *Economic Botany* 44:6–27.

———. 2001. "George Beadle's Other Hypothesis: One-Gene, One-Trait." *Genetics* 158:487–493.

———. 2003. "The Morphology of Maize and Teosinte." Laboratory of Genetics, University of Wisconsin-Madison. http://teosinte.wisc.edu/morphology .html, accessed April 20, 2008.

———. 2004. "The Genetics of Maize Evolution." *Annual Review of Genetics* 38:37–59.

Doebley, John, Brandon S. Gaut, and Bruce D. Smith. 2006. "The Molecular Genetics of Crop Domestication." *Cell* 127 (7):1309–1321.

Doebley, John, Major M. Goodman, and C. W. Stuber. 1984. "Isoenzymatic Variation in *Zea* (Gramineae)." *Systematic Botany* 9:203–218.

Dorweiler, A., A. Stec, J. Kermicle, and J. Doebley. 1993. "Teosinte Glume Architecture 1: A Genetic Locus Controlling a Key Step in Maize Evolution." *Science* 262 (5131):233–235.

Douglas, Mary, ed. 1984. *Food in the Social Order: Studies of Food and Festivities in Three American Communities.* New York: Russell Sage Foundation.

Downard, Kevin M. 2007. "Historical Account: Francis William Aston: The Man behind the Mass Spectrograph." *European Journal of Mass Spectrometry* 13 (3):177–190.

Duffy, Lisa G. 2011. "Maize and Stone: A Functional Analysis of the Manos and Metates of Santa Rita Corozal, Belize." Master's thesis, University of Central Florida.

Dull, Robert A. 2006. "The Maize Revolution: A View from El Salvador." In *Histories of Maize: Multidisciplinary Approaches to the Prehistory, Biogeography, Domestication, and Evolution of Maize,* edited by J.E. Staller, R.H. Tykot, and B.F. Benz, pp. 357–365. Amsterdam: Academic Press.

Ekdahl, Erik J., Jane L. Teranes, Thomas P. Guilderson, Charles L. Turton, John H. McAndrews, Chad A. Wittkop, et al. 2004. "Prehistorical Record of Cultural Eutrophication from Crawford Lake, Canada." *Geology* 32 (9):745–748.

Emshwiller, Eve. 2006. "Genetic Data and Plant Domestication." In *Documenting Domestication: New Genetic and Archaeological Paradigms,* edited by M.A. Zeder, D.G. Bradley, E. Emshwiller, and B.D. Smith, pp. 99–122. Berkeley: University of California Press.

Engels, Friedrich. 1972. *Origin of the Family, Private Property and the State.* New York: Pathfinder Press.

Erickson, C.L., and W. Balée. 2006. "The Historical Ecology of a Complex Landscape in Bolivia." In *Time and Complexity in Historical Ecology: Studies in the Neotropical Lowlands,* edited by W. Balée and C.L. Erickson, pp. 187–234. New York: Columbia University Press.

Ericson, J.E., M. West, C.H. Sullivan, and H.W. Krueger. 1989. "The Development of Maize Agriculture in the Viru Valley, Peru." In *The Chemistry of Prehistoric Human Bone,* edited by T.D. Price, pp. 68–104. Cambridge, UK: Cambridge University Press.

Eubanks, Mary. 1997. "Reevaluation of the Identification of Ancient Maize Pollen from Alabama." *American Antiquity* 62 (1):139–145.

———. 1999. *Corn in Clay: Maize Paleoethnobotany in Pre-Columbian Art.* Gainesville: University Press of Florida.

———. 2001a. "The Mysterious Origin of Maize." *Economic Botany* 55:492–514.

———. 2001b. "An Interdisciplinary Perspective on the Origin of Maize." *Latin American Antiquity* 12 (1):91–98.

Evans, Susan Toby. 2013. *Ancient Mexico and Central America: Archaeology and Culture History.* 3rd ed. London: Thames and Hudson.

Faegri, Knut, Johannes Iversen, Peter Emil Kaland, and Knut Krzywinski. 1989. *Textbook of Pollen Analysis.* 4th ed. Caldwell, NJ: Blackburn Press.

Fearn, Miriam L., and Kam-Biu Liu. 1995. "Maize Pollen of 3500 B.P. from Southern Alabama." *American Antiquity* 60 (1):109–117.

Feddema, Vicki L. 1993. "Early Formative Subsistence and Agriculture in Southeastern Mesoamerica." Master's thesis, University of British Columbia.

Fish, Susanne K. 2004. "Corn, Crops, and Cultivation in the North American Southwest." In *People and Plants in Ancient Western North America*, edited by P. E. Minnis, pp. 115–166. Washington: Smithsonian Books.

Flannery, Kent V., ed. 1986a. *Guilá Naquitz: Archaic Foraging and Early Agriculture in Oaxaca, Mexico*. Orlando: Academic Press.

———. 1986b. "The Quantification of Subsistence Data: An Introduction to Part V." In *Guilá Naquitz: Archaic Foraging and Early Agriculture in Oaxaca, Mexico,* edited by K. V. Flannery, pp. 249–253. Orlando: Academic Press.

———. 1986c. "The Research Problem." In *Guilá Naquitz: Archaic Foraging and Early Agriculture in Oaxaca, Mexico,* edited by K. V. Flannery, pp. 3–18. Orlando: Academic Press.

———. 2002. "Editorial Comment." *Antiquity* 76:289–291.

Flannery, Kent V., and Joyce Marcus. 2001. *Richard Stockton MacNeish, 1918–2001: A Biographical Memoir*. Washington, DC: The National Academy Press.

Fournier, Patricia. 1998. "El complejo nixtamal/comal/tortilla en Mesoamérica." *Boletín de Antropología Americana* 32:13.

French, David H. 1971. "An Experiment in Water-Sieving." *Anatolian Studies* 21:59–64.

Fritz, Gayle J. 1994. "Are the First American Farmers Getting Younger?" *Current Anthropology* 35:305–309.

Fukunaga, Kenji, Jason Hill, Yves Vigouroux, Yoshihiro Matsuoka, Jesus Sanchez G., Kejun Liu, et al. 2005. "Genetic Diversity and Population Structure of Teosinte." *Genetics* 169:2241–2254.

Gamwell, Lynn. 2003. "Perceptions of Science: Beyond the Visible-Microscopy, Nature, and Art." *Science* 299 (5603):49–50.

GHR. 2015. Genetics Home Reference. Bethesda, MD: National Library of Medicine, National Institutes of Health. www.ghr.nlm.nih.gov/, accessed March 4, 2015.

Girard, Raphael. 1979. *Esotericism of the Popol Vuh*. Translated by B. A. Moffett. Pasadena, CA: Theosophical University Press. Online ed., www.theosociety.org/pasadena/popolvuh/pv-hp.htm.

Goldberg, R. B., T. P. Beals, and P. M. Sanders. 1993. "Anther Development: Basic Principles and Practical Applications." *Plant Cell* 5:1217–1229.

Goldemberg, José, Suani Teixeira Coelho, Plinio Mário Nastari, and Oswaldo Lucon. 2004. "Ethanol Learning Curve—the Brazilian Experience." *Biomass and Bioenergy* 26:301–304.

Goloubinoff, Pierre, Svante Pääbo, and Allan C. Wilson. 1993. "Evolution of Maize Inferred from Sequence Diversity of an *Adh2* Gene Segment from Archaeological Specimens." *Proceedings of the National Academy of Sciences, USA* 90:1997–2001.

Gore, Michael A., Jer-Ming Chia, Robert J. Elshire, Qi Sun, Elhan S. Ersoz, Bonnie L. Hurwitz, et al. 2009. "A First-Generation Haplotype Map of Maize." *Science* 326 (5956):1115–1117.

Green, Richard E., Johannes Krause, Susan E. Ptak, Adrian W. Briggs, Michael T. Ronan, Jan F. Simons, et al. 2006. "Analysis of One Million Base Pairs of Neanderthal DNA." *Nature* 444:330–336.

Grobman, Alexander, and Duccio Bonavia. 1978. "Pre-Ceramic Maize on the North-Central Coast of Peru." *Nature* 276 (5686):386–387.

Grobman, Alexander, Ducci Bonavia, David H. Kelley, Paul C. Mangelsdorf, and Julián Cámara-Hernández. 1977. "Study of Pre-Ceramic Maize from Huarmey, North Central Coast of Peru." *Botanical Museum Leaflets, Harvard University* 25 (8):221–242.

Grobman, Alexander, Wilfredo Salhuana, Ricardo Sevilla, and Paul C. Mangelsdorf. 1961. *Races of Maize in Peru: Their Origins, Evolution and Classification. Publication 915.* Washington, DC: National Academy of Science and National Research Council.

Haas, Jonathan, and Winifred Creamer. 2006. "Crucible of Andean Civilization: The Peruvian Coast from 3000 to 1800 BC." *Current Anthropology* 47 (5):745–775.

Hart, John P. 2008. "Evolving the Three Sisters: The Changing Histories of Maize, Bean, and Squash in New York and the Greater Northeast." In *Current Northeast Paleoethnobotany II,* edited by John P. Hart, pp. 87–99. New York State Museum Bulletin 512. Albany: State University of New York.

———. 2014. "A Critical Assessment of Current Approaches to Investigations of the Timing, Rate, and Adoption Trajectories of Domesticates in the Midwest and Great Lakes." *Midwest Archaeological Conference Inc. Occasional Papers* 1:161–174.

Hart, John P. and R. G. Matson. 2009. "The Use of Multiple Discriminant Analysis in Classifying Prehistoric Phytolith Assemblages Recovered from Cooking Residues." *Journal of Archaeological Science* 36:74–83.

Hart, John P., R. G. Matson, Robert G. Thompson, and Michael Blake. 2011. "Teosinte Inflorescence Phytolith Assemblages Mirror *Zea* Taxonomy." *PLoS ONE* 6 (3):e18349.

Hayden, Brian. 1990. "Nimrods, Piscators, Pluckers and Planters: The Emergence of Food Production." *Journal of Anthropological Archaeology* 9: 31–69.

———. 2009. "The Proof Is in the Pudding: Feasting and the Origins of Domestication." *Current Anthropology* 50:597–601.

Hays-Gilpin, Kelley, and Michelle Hegmon. 2005. "The Art of Ethnobotany: Depictions of Maize and Other Plants in the Prehispanic Southwest." In *Engaged Anthropology: Research Essays on North American Archaeology, Ethnobotany, and Museology,* edited by M. Hegmon and B. S. Eiselt, pp. 89–113. Anthropological Papers, no. 94. Ann Arbor: University of Michigan Museum of Anthropology.

Heckenberger, Michael J. 2005. *The Ecology of Power: Culture, Place, and Personhood in the Southern Amazon, A.D. 1000–2000.* New York: Routledge.

Hill, Jane H. 2001. "Proto-Uto-Aztecan: A Community of Cultivators in Central Mexico?" *American Anthropologist* 103 (4):913–934.

Horn, Sally P. 2006. "Pre-Columbian Maize Agriculture in Costa Rica: Pollen and Other Evidence from Lake and Swamp Sediments." In *Histories of Maize: Multidisciplinary Approaches to the Prehistory, Biogeography, Domestication, and Evolution of Maize,* edited by J. E. Staller, R. H. Tykot, and B. F. Benz, pp. 367–380. Amsterdam: Academic Press.

Horowitz, Norman. 1990. *George Wells Beadle, 1903–1989: A Biographical Memoir.* Washington, DC: National Academy of Sciences.

Horton, H. Robert, Laurence A. Moran, K. Gray Scrimgeour, Marc D. Perry and J. David Rawn. 2006. *Principles of Biochemistry.* 4th ed. Upper Saddle River, NJ: Pearson Prentice Hall.

Huber, E. K., and C. R. Van West, eds. 2005. *Fence Lake Project.* Tempe: Statistical Research.

Hudson, Charles. 1984. *Elements of Southeastern Indian Religion.* Leiden: E. J. Brill.

Hufford, Matthew B., Xun Xu, Joost van Heerwaarden, Tanja Pyhajarvi, Jer-Ming Chia, Reed A. Cartwright, et al. 2012. "Comparative Population Genomics of Maize Domestication and Improvement." *Nature Genetics* 44 (7):808–811.

Iltis, Hugh H. 1983. "From Teosinte to Maize: The Catastrophic Sexual Transmutation." *Science* 222 (4626):886–894.

———. 2000. "Homeotic Sexual Translocations and the Origin of Maize (*Zea mays,* Poaceae): A New Look at an Old Problem." *Economic Botany* 54 (1):7–42.

———. 2006. "Origin of Polystichy in Maize." In *Histories of Maize: Multidisciplinary Approaches to the Prehistory, Biogeography, Domestication, and Evolution of Maize,* edited by J. E. Staller, R. H. Tykot, and B. F. Benz, pp. 21–53. Amsterdam: Academic Press.

Iltis, Hugh H., and John F. Doebley. 1980. "Taxonomy of *Zea* (Gramineae). II. Sub-Specific Categories in the *Zea mays* Complex and a Generic Synopsis." *American Journal of Botany* 67:994–1004.

Jacobs, Bonnie F., John D. Kingston, and Louis L. Jacobs. 1999. "The Origin of Grass-Dominated Ecosystems." *Annals of the Missouri Botanical Garden* 86 (2):590–643.

Jaenicke-Després, Viviane, Ed S. Buckler, Bruce D. Smith, M. Thomas, P. Gilbert, Alan Cooper, et al. 2003. "Early Allelic Selection in Maize as Revealed by Ancient DNA." *Science* 302 (5648):1206–1208.

Jaenicke-Després, Viviane, and Bruce D. Smith. 2006. "Ancient DNA and the Integration of Archaeological and Genetic Approaches to the Study of Maize Domestication." In *Histories of Maize: Multidisciplinary Approaches to the Prehistory, Biogeography, Domestication, and Evolution of Maize,* edited by J. E. Staller, R. H. Tykot, and B. F. Benz, pp. 83–96. Amsterdam: Academic Press.

Johnson, Frederick, ed. 1972. *The Prehistory of the Tehuacán Valley.* Vol. 4: *Chronology and Irrigation.* Austin: University of Texas Press.

Johnson, Frederick, and Richard S. MacNeish. 1972. "Chronometric Dating." In *The Prehistory of the Tehuacán Valley*. Vol. 4: *Chronology and Irrigation*, edited by F. Johnson, pp. 3–55. Austin: University of Texas Press.

Joralemon, Peter David. 1971. *A Study of Olmec Iconography*. Washington, DC: Dumbarton Oaks.

Karttunen, Frances. 1992. *An Analytical Dictionary of Nahuatl*. Norman: University of Oklahoma Press.

Katiyar, S. K., and J. K. S. Sachan. 1992. "Scanning Electron Microscopic Studies of Pollen Grains in the Tribe Maydeae." *Maize Genetics Cooperation Newsletter* 66:91–92.

Katz, S. H., M. L. Hediger, and L. A. Valleroy. 1974. "Traditional Maize Processing Techniques in the New World." *Science* 184 (4138):765–773.

Katzenberg, M. Anne. 2003. "Comment on John Smalley and Michael Blake, Sweet Beginnings: Stalk Sugar and the Domestication of Maize." *Current Anthropology* 44 (5):675–703.

Keeley, Jon E., and Philip W. Rundel. 2003. "Evolution of CAM and C4 Carbon◌Concentrating Mechanisms." *International Journal of Plant Sciences* 164 (3 Suppl.):S55–S77.

Kelley, David H., and Duccio Bonavia. 1963. "New Evidence for Preceramic Maize on the Coast of Peru." *Ñawpa Pacha* 1:39–41.

Killion, Thomas W. 2013. "Nonagricultural Cultivation and Social Complexity: The Olmec, Their Ancestors, and Mexico's Southern Gulf Coast Lowlands." *Current Anthropology* 54 (5):569–606.

Kirkby, Anne V. T. 1973. *The Use of Land and Water Resources in the Past and Present Valley of Oaxaca*. Prehistory and Human Ecology of the Valley of Oaxaca, vol. 1. Memoirs of the University of Michigan Museum of Anthropology, no. 5. Ann Arbor: University of Michigan.

Knight, Vernon James, Jr. 1986. "The Institutional Organization of Mississippian Religion." *American Antiquity* 51 (4):675–687.

Kroeber, Alfred L. 1933. *Anthropology*. New York: Harcourt Brace and Co.

Lanning, Edward P. 1963. "Olmec and Chavin: Reply to Michael D. Coe." *American Antiquity* 29 (1):99–101.

Largy, Tonya, and E. Pierre Morenon. 2008. "Maize Agriculture in Coastal Rhode Island: Imaginative, Illusive or Intensive?" In *Current Northeast Paleoethnobotany II*, edited by John P. Hart, pp. 73–86. New York State Museum Bulletin 512. Albany: University of the State of New York.

Lauter, Nick, and John Doebley. 2002. "Genetic Variation for Phenotypically Invariant Traits Detected in Teosinte: Implications for the Evolution of Novel Forms." *Genetics* 160:333–342.

Lazaridis, Iosif, Nick Patterson, Alissa Mittnik, Gabriel Renaud, Swapan Mallick, Karola Kirsanow, et al. 2014. "Ancient Human Genomes Suggest Three Ancestral Populations for Present-Day Europeans." *Nature* 513 (7518):409–413.

LeBlanc, Steven A., Lori S. Cobb Kreisman, Brian M. Kemp, Francis E. Smiley, Shawn W. Carlyle, Anna N. Dhody, et al. 2007. "Quids and Aprons: Ancient DNA from Artifacts from the American Southwest." *Journal of Field Archaeology* 32 (2):161–175.

Lesure, Richard, ed. 2011. *Early Mesoamerican Social Transformations: Archaic and Formative Lifeways in the Soconusco Region*. Berkeley: University of California Press.

Lévi-Strauss, Claude. 1969. *The Raw and the Cooked*. Translated by John and Doreen Weightman. New York: Harper and Row.

Libby, Willard F. 1951. "Radiocarbon Dates, II." *Science* 114 (2960):291–296.

———. 1952. "Chicago Radiocarbon Dates, III." *Science* 116 (3025):673–681.

Libby, Willard F., Ernest C. Anderson, and James R. Arnold. 1949. "Age Determination by Radiocarbon Content: World-Wide Assay of Natural Radiocarbon." *Science* 109 (2827):227–228.

Long, Austin, Bruce F. Benz, D. J. Donahue, A. J. T. Jull, and L. J. Toolin. 1989. "First Direct AMS Dates on Early Maize from Tehuacán, Mexico." *Radiocarbon* 31 (3):1035–1040.

Loy, Thomas H., Mathew Spriggs, and Stephen Wickler. 1992. "Direct Evidence for Human Use of Plants 28,000 Years Ago: Starch Residues on Stone Artefacts from the Northern Solomon Islands." *Antiquity* 66:898–912.

Luna V., S., J. Figueroa M., B. Baltazar M., R. Gomez L., R. Townsend, and J. B. Schoper. 2001. "Maize Pollen Longevity and Distance Isolation Requirements for Effective Pollen Control." *Crop Science* 41:1551–1557.

Mabry, Jonathan B., ed. 2008. *Las Capas: Early Irrigation and Sedentism in a Southwestern Floodplain*. Tucson: Center for Desert Archaeology.

MacNeish, Richard S. 1958. *Preliminary Archaeological Investigations in the Sierra de Tamaulipas, Mexico*. Transactions of the American Philosophical Society, vol. 48, part 6. Philadelphia: American Philosophical Society.

MacNeish, Richard S., and Mary W. Eubanks. 2000. "Comparative Analysis of the Río Balsas and Tehuacán Models for the Origin of Maize." *Latin American Antiquity* 11:3–20.

MacNeish, Richard S., Melvin L. Fowler, Angel García Cook, Frederick A. Peterson, Antoinette Nelken-Terner, and James A. Neely, eds. 1972. *The Prehistory of the Tehuacán Valley*. Vol. 5: *Excavations and Reconnaissance*. Austin: University of Texas Press.

MacNeish, Richard S., Antoinette Nelken-Terner, and Irmgard W. Johnson, eds. 1967. *The Prehistory of the Tehuacán Valley*. Vol. 2: *Nonceramic Artifacts*. Austin: University of Texas Press.

MacNeish, Richard S., and Frederick A. Peterson. 1962. *The Santa Marta Rock Shelter, Ocozocoautla, Chiapas, Mexico*. Papers of the New World Archaeological Foundation, no. 14. Provo, UT: Brigham Young University.

MacNeish, Richard S., Frederick A. Peterson, and Kent V. Flannery, eds. 1970. *The Prehistory of the Tehuacán Valley*. Vol. 3: *Ceramics*. Austin: University of Texas Press.

Mangelsdorf, Paul C. 1974. *Corn: Its Origin, Evolution and Improvement*. Cambridge, MA: Harvard University Press.

———. 1983. "The Mystery of Corn: New Perspectives." *Proceedings of the American Philosophical Society* 127:215–247.

———. 1986. "The Origin of Corn." *Scientific American* 255:80–86.

Mangelsdorf, Paul C., Richard S. MacNeish, and Walton C. Galinat. 1964. "Domestication of Corn." *Science* 143 (3612):538–545.

———. 1967. "Prehistoric Wild and Cultivated Maize." In *The Prehistory of the Tehuacan Valley.* Vol. 1: *Environment and Subsistence,* edited by D.S. Byers, pp. 178–200. Austin: University of Texas Press.

Mangelsdorf, Paul C., and R.G. Reeves. 1938. "The Origin of Maize." *Proceedings of the National Academy of Sciences, USA* 24 (8):303–312.

———. 1939. "The Origin of Indian Corn and Its Relatives." *Texas Agricultural Experiment Station, Bulletin* 574:1–315.

Mangelsdorf, Paul C., and C. Earle Smith. 1949. "New Archaeological Evidence on Evolution in Maize." *Harvard University Botanical Museum Leaflets* 13:213–247.

Marcus, Joyce, and Kent V. Flannery. 1996. *Zapotec Civilization: How Urban Society Evolved in Mexico's Oaxaca Valley.* New York: Thames and Hudson.

Martínez Donjuán, Guadalupe. 1985. "El Sitio Olmeca de Teopantecuanitlán En Guerrero." *Anales de Antropología* 22:216–226.

———. 1994. "Los Olmecas En El Estado de Guerrero." In *Los Olmecas En Mesoamérica,* edited by J.E. Clark, pp. 143–163. México, DF: El Equilibrista.

Marx, Karl. 1965. *Pre-Capitalist Economic Formations.* Translated by J. Cohen. New York: International Publishers.

Matson, R.G. 1991. *The Origins of Southwestern Agriculture.* Tucson: University of Arizona Press.

———. 2003. "The Spread of Maize Agriculture into the U.S. Southwest." In *Examining the Farming/Language Dispersal Hypothesis,* edited by P. Bellwood and C. Renfrew, pp. 341–356. Cambridge, UK: McDonald Institute for Archaeological Research, University of Cambridge.

Matsuoka, Yoshihiro, Yves Vigouroux, Major M. Goodman, Jesus Sanchez G., Edward Buckler, and John Doebley. 2002. "A Single Domestication for Maize Shown by Multilocus Microsatellite Genotyping." *Proceedings of the National Academy of Sciences, USA* 99 (9):6080–6084.

Mauseth, James D. 2008. *Botany: An Introduction to Plant Biology.* Sudbury, MA: Jones and Bartlett. www.biology.jbpub.com/botany/4e/glossary.cfm, accessed March 4, 2015.

McCann, James C. 2005. *Maize and Grace: Africa's Encounter with a New World Crop, 1500–2000.* Cambridge, MA: Harvard University Press.

McClung de Tapia, Emily. 2000. "Prehispanic Agricultural Systems in the Basin of Mexico." In *Imperfect Balance: Landscape Transformations in the Precolumbian Americas,* edited by D.L. Lentz, pp. 121–146. New York: Columbia University Press.

McCormick, S. 1993. "Male Gametophyte Development." *Plant Cell* 5:1265–1275.

McGovern, Patrick E. 2009. *Uncorking the Past: The Quest for Wine, Beer, and Other Alcoholic Beverages.* Berkeley: University of California Press.

Mejía, Danilo. 2003. *Maize: Post-Harvest Operation.* Rome: Food and Agriculture Organization of the United Nations. www.fao.org/inpho/inpho-post-harvest-compendium/cereals-grains/en/

Mellars, Paul. 2006. "Why Did Modern Human Populations Disperse from Africa ca. 60,000 Years Ago?" *Proceedings of the National Academy of Sciences, USA* 103 (25):9381–9386.

Merrill, William L., Robert J. Hard, Jonathan B. Mabry, Gayle J. Fritz, Karen R. Adams, John R. Roney, et al. 2009. "The Diffusion of Maize to the Southwestern United States and Its Impact." *Proceedings of the National Academy of Sciences, USA* 106 (50):21019–21026.

Minnis, Paul E. 1984. "Macroplant Remains." In *The Galaz Ruin: A Prehistoric Village in Soutwestern New Mexico,* edited by R. Anyon and S. A. LeBlanc, pp. 193–200. Albuquerque: Maxwell Museum of Anthropology and University of New Mexico Press.

———. 1985. *Social Adaptation to Food Stress.* Chicago: University of Chicago Press.

Minnis, Paul E., and Steven A. LeBlanc. 1976. "An Efficient, Inexpensive Arid Lands Flotation System." *American Antiquity* 46:491–493.

Moerman, Daniel E. 1998. *Native American Ethnobotany.* Portland, OR: Timber Press.

Moreiras Reynaga, Diana K. 2013. "Pre-Columbian Diets in the Soconusco Revisited: A Dietary Study through Stable Isotopic Analysis." Master's thesis, University of British Columbia.

Morgan, Lewis Henry. 1877. *Ancient Society, or Researches in the Lines of Human Progress from Savagery through Barbarism to Civilization.* New York: Henry Holt.

Moseley, Michael. 1975. *Maritime Foundations of Andean Civilization.* Menlo Park: Cummings.

———. 1992. "Maritime Foundations and Multilinear Evolution: Retrospect and Prospect." *Andean Past* 3:5–42.

———. 2001. *The Incas and Their Ancestors: The Archaeology of Peru.* New York: Thames and Hudson.

Moseley, Michael E., Donna J. Nash, Patrick Ryan Williams, Susan D. deFrance, Ana Miranda, and Mario Ruales. 2005. "Burning Down the Brewery: Establishing and Evacuating an Ancient Imperial Colony at Cerro Baúl, Peru." *Proceedings of the National Academy of Sciences, USA* 102 (48):17264–17271.

Mulholland, S. C., and C. Prior. 1993. "AMS Radiocarbon Dating of Phytoliths." In *Current Reserch in Phytolith Analysis: Applications in Archaeology and Paleoecology,* edited by D. M. Pearsall and D. R. Piperno, pp. 21–23. MASCA Research Papers in Science and Archaeology, vol. 10. Philadelphia: Museum of Archaeology and Anthropology, University of Pennsylvania.

Myers, Thomas P. 2006. "Hominy Technology and the Emergence of Mississippian Societies." In *Histories of Maize: Multidisciplinary Approaches to the Prehistory, Biogeography, Domestication, and Evolution of Maize,* edited by J. E. Staller, R. H. Tykot, and B. F. Benz, pp. 511–520. Amsterdam: Academic Press.

Neff, Hector, Deborah M. Pearsall, John G. Jones, Bárbara Arroyo, Shawn K. Collins, and Dorothy E. Freidel. 2006a. "Early Maya Adaptive Patterns: Mid-Late Holocene Paleoenvironmental Evidence from Pacific Guatemala." *Latin American Antiquity* 17 (3):287–315.

Neff, Hector, Deborah R. Pearsall, John G. Jones, Bárbara Arroyo, and Dorothy E. Freidel. 2006b. "Climate Change and Population History in the

Pacific Lowlands of Southern Mesoamerica." *Quaternary Research* 65:390–400.

Nelson, D. Erle, R. G. Korteling, and W. R. Stott. 1977. "Carbon-14: Direct Detection at Natural Concentrations." *Science* 198 (4316):507–508.

NHGRI. 2014. Talking Glossary of Genetic Terms. Bethesda, MD: National Human Genome Research Institute. www.genome.gov/10002096, accessed April 8, 2007.

NNDC-BNL. 2015. Interactive Chart of Nuclides. Upton, NY: National Nuclear Data Center—Brookhaven National Laboratory. www.nndc.bnl.gov/chart/reCenter.jsp?z=6&n=6, accessed March 4, 2015.

O'Leary, Marion H. 1988. "Carbon Isotopes in Photosynthesis." *BioScience* 38 (5):328–336.

Pääbo, Svante, Hendrik Poinar, David Serre, Viviane Jaenicke-Després, Juliane Hebler, Nadin Rohland, et al. 2004. "Genetic Analyses from Ancient DNA." *Annual Review of Genetics* 38 (1):645–679.

Panzea. 2014. Panzea: The Maize Diversity Project. www.panzea.org, accessed July 14, 2014.

Papas, Rebecca K., John E. Sidle, Emmanuel S. Wamalwa, Thomas O. Okumu, Kendall L. Bryant, Joseph L. Goulet, et al. 2010. "Estimating Alcohol Content of Traditional Brew in Western Kenya Using Culturally Relevant Methods: The Case for Cost over Volume." *AIDS and Behavior* 14 (4):836–844.

Pattberg, Philip. 2007. "Conquest, Domination and Control: Europe's Mastery of Nature in Historic Perspective." *Journal of Political Ecology* 14:1–9.

Pearsall, Deborah M. 1978. "Phytolith Analysis of Archaeological Soils: Evidence for Maize Cultivation in Formative Ecuador." *Science* 199 (4325):177–178.

———. 1988. "Interpreting the Meaning of Macroremain Abundance: The Impact of Source and Context." In *Current Paleoethnobotany: Analytical Methods and Cultural Interpretations of Archaeological Plant Remains,* edited by C. A. Hastorf and V. S. Popper, pp. 97–118. Chicago: University of Chicago Press.

———. 1995. "'Doing' Paleoethnobotany in the Tropical Lowlands: Adaptation and Innovation in Methodology." In *Archaeology in the Lowland American Tropics: Current Analytical Methods and Recent Applications,* edited by P. W. Stahl, pp. 113–129. Cambridge UK: Cambridge University Press.

———. 2000. *Paleoethnobotany: A Handbook of Procedures.* 2nd ed. San Diego: Academic Press.

———. 2002. "Maize Is Still Ancient in Prehistoric Ecuador: The View from Real Alto, with Comments on Staller and Thompson." *Journal of Archaeological Science* 29 (1):51–55.

———. 2003. "Plant Food Resources of the Ecuadorian Formative: An Overview and Comparison to the Central Andes." In *Archaeology of Formative Ecuador: A Symposium at Dumbarton Oaks, 7 and 8 October 1995,* edited by J. S. Raymond and R. L. Burger, pp. 213–257. Washington, DC: Dumbarton Oaks Research Library and Collection.

Pearsall, Deborah M., Karol Chandler-Ezell, and James A. Zeidler. 2004. "Maize in Ancient Ecuador: Results of Residue Analysis of Stone Tools from the Real Alto Site." *Journal of Archaeological Science* 31 (4):423–442.

Pearsall, Deborah M., and Dolores R. Piperno. 1990. "Antiquity of Maize Cultivation in Ecuador: Summary and Reevaluation of the Evidence." *American Antiquity* 55 (2):324–337.

Perry, Linda, Ruth Dickau, Sonia Zarrillo, Irene Holst, Deborah M. Pearsall, Dolores R. Piperno, et al. 2007. "Starch Fossils and the Domestication and Dispersal of Chili Peppers (*Capsicum* spp. L.) in the Americas." *Science* 315 (5814):986–988.

Perry, Linda, Daniel H. Sandweiss, Dolores R. Piperno, Kurt Dademaker, Michael A. Malpass, Adán Umire, et al. 2006. "Early Maize Agriculture and Interzonal Interaction in Southern Peru." *Nature* 440:76–79.

Piperno, Dolores R. 2003. "A Few Kernels Short of a Cob: On the Staller and Thompson Late Entry Scenario for the Introduction of Maize into Northern South America." *Journal of Archaeological Science* 30:831–836.

———. 2006. *Phytoliths: A Comprehensive Guide for Archaeologists and Paleoecologists*. Walnut Creek, CA: AltaMira Press.

Piperno, Dolores R., Karen H. Clary, Richard G. Cooke, Anthony J. Ranere, and D. Weiland. 1985. "Preceramic Maize in Central Panama: Phytolith and Pollen Evidence." *American Anthropologist* 87 (4):871–878.

Piperno, Dolores R., and Kent V. Flannery. 2001. "The Earliest Archaeological Maize (*Zea mays* L.) from Highland Mexico: New Accelerator Mass Spectrometry Dates and Their Implications." *Proceedings of the National Academy of Sciences, USA* 98 (4):2101–2103.

Piperno, Dolores R., and Deborah M. Pearsall. 1998. *The Origins of Agriculture in the Lowland Neotropics*. San Diego: Academic Press.

Piperno, Dolores R., Anthony J. Ranere, Irene Holst, and Patricia Hansell. 2000. "Starch Grains Reveal Early Root Crop Horticulture in the Panamanian Tropical Forest." *Nature* 407:894–897.

Piperno, Dolores R., Anthony J. Ranere, Irene Holst, Jose Iriarte, and Ruth Dickau. 2009. "Starch Grain and Phytolith Evidence for Early Ninth Millennium B.P. Maize from the Central Balsas River Valley, Mexico." *Proceedings of the National Academy of Sciences, USA* 106 (13):5019–5024.

Pluciennik, Mark. 2001. "Archaeology, Anthropology and Subsistence." *Journal of the Royal Anthropological Institute* 7 (4):741–758.

Pohl, Mary E. D., Dolores R. Piperno, Kevin O. Pope, and John G. Jones. 2007. "Microfossil Evidence for Pre-Columbian Maize Dispersals in the Neotropics from San Andrés, Tabasco, Mexico." *Proceedings of the National Academy of Sciences, USA* 104 (16):6870–6875.

Pollan, Michael. 2001. *The Botany of Desire: A Plant's-Eye View of the World*. New York: Random House.

———. 2006. *The Omnivore's Dilemma: A Natural History of Four Meals*. New York: The Penguin Press.

Pollard, Helen Perlstein. 1997. "Recent Research in West Mexican Archaeology." *Journal of Archaeological Research* 5 (4):345–384.

Pope, Kevin O., Mary D. Pohl, John G. Jones, D. L. Lentz, Christopher von Nagy, F. J. Vega, et al. 2001. "Origin and Environmental Setting of Ancient Agriculture in the Lowlands of Mesoamerica." *Science* 292 (5520):1370–1373.

Popper, Virginia S. 1982 "Restos Botánicos: Analisis general de las muestras." In *Los Gavilanes: Mar, desierto y oasis en la historia del hombre*, edited by Duccio Bonavia, pp. 148–156. Lima: Corporación Financiera de Desarollo S.A and Instituto Arqueológico Alemán.

Prasad, Vandana, Caroline A. E. Strömberg, Habib Alimohammadian, and Ashok Sahni. 2005. "Dinosaur Coprolites and the Early Evolution of Grasses and Grazers." *Science* 310 (5751):1177–1180.

Przeworski, M. 2003. "Estimating the Time since the Fixation of a Beneficial Allele." *Genetics* 164:1667–1676.

Purugganan, Michael D., and Dorian Q. Fuller. 2009. "The Nature of Selection during Plant Domestication." *Nature* 457 (7231):843–848.

Quilter, Jeffrey, and Terry Stocker. 1983. "Subsistence Economies and the Origins of Andean Complex Societies." *American Anthropologist* 85:545–562.

Ranere, Anthony J., Dolores R. Piperno, Irene Holst, Ruth Dickau, and José Iriarte. 2009. "The Cultural and Chronological Context of Early Holocene Maize and Squash Domestication in the Central Balsas River Valley, Mexico." *Proceedings of the National Academy of Sciences, USA* 106 (13):5014–5018.

Raymond, J. Scott. 1981. "The Maritime Foundations of Andean Civilization: A Reconsideration of the Evidence." *American Antiquity* 45:806–821.

Recinos, Adrián, Delia Goetz, and Sylvanus G. Morley. 1950. *Popol Vuh*. Norman: University of Oklahoma Press.

Reichert, Edward T. 1913. *Differentiation and Specificity of Starches in Relation to Genera, Species, etc.; Stereochemistry Applied to Protoplasmic Processes and Products, and as a Strictly Scientific Basis for the Classification of Plants and Animals*. Washington, DC: Carnegie Institution of Washington.

Rice, Prudence M. 2005. *Pottery Analysis: A Sourcebook*. Chicago: University of Chicago Press.

Richards, Michael P., and Erik Trinkaus. 2009. "Isotopic Evidence for the Diets of European Neanderthals and Early Modern Humans." *Proceedings of the National Academy of Sciences, USA* 106 (38):16034–16039.

Rodríguez Martínez, Ma. del Carmen, Ponciano Ortíz Ceballos, Michael D. Coe, Richard A. Diehl, Stephen D. Houston, Karl A. Taube, et al. 2006. "Oldest Writing in the New World." *Science* 313 (5793):1610–1614.

Rosenswig, Robert M. 2010. *The Beginnings of Mesoamerican Civilization: Inter-Regional Interaction and the Olmec*. Cambridge, UK: Cambridge University Press.

Rowley-Conwy, Peter. 2011. "Westward Ho! The Spread of Agriculturalism from Central Europe to the Atlantic." *Current Anthropology* 52 (S4):S431-S451.

Sahagún, Bernardino de. 1950–1982. *Florentine Codex: General History of the Things of New Spain*. Translated by C. E. Dibble and A. J. O. Anderson. Salt Lake City: University of Utah Press.

Sanders, William T., and Joseph Marino. 1970. *New World Prehistory: Archaeology of the American Indian*. Englewood Cliffs, NJ: Prentice Hall.

Sanderson, Katharine. 2005. "Powerful Promise for Grass That's as High as an Elephant's Eye." *Chemistry World* (online journal of the Royal Society of

Chemistry). www.rsc.org/chemistryworld/News/2005/September/09090502 .asp, accessed March 20, 2005.

Saturno, William A., David Stuart, and Boris Beltrán. 2006. "Early Maya Writing at San Bartolo, Guatemala." *Science* 311 (5765):1281–1283.

Saturno, William A., Karl Taube, David E. Stuart, and Heather Hurst. 2005. *The Murals of San Bartolo, El Petén, Guatemala: Part I: The North Wall.* Barnardsville, NC: Center for Ancient American Studies.

Sauer, Carl O. 1952. *Agricultural Origins and Dispersals.* New York: American Geographical Society.

Scheffler, Timothy E. 2008. "The El Gigante Rockshelter, Honduras." PhD diss., Pennsylvania State University.

Schultheis, Jonathan R. 1998. "Sweet Corn Production." Raleigh, NC: Department of Horticultural Science, North Carolina State University. www.ces .ncsu.edu/depts/hort/hil/hil-13.html, accessed July 14, 2012.

Scitable. 2015. www.nature.com/scitable. Cambridge, MA: Nature Publishing Group. Accessed March 4, 2015.

Shady, Ruth. 2006. "Caral-Supe and the North-Central Area of Peru: The History of Maize in the Land Where Civilization Came into Being." In *Histories of Maize: Multidisciplinary Approaches to the Prehistory, Biogeography, Domestication, and Evolution of Maize,* edited by J. E. Staller, R. H. Tykot, and B. F. Benz, pp. 381–402. Amsterdam: Academic Press.

Shady Solís, Ruth, and Carlos Leyva, eds. 2003. *La ciudad sagrada de Caral-Supe: Los orígenes de la civilización andina y la formación del Estado prístino en el antiguo Perú.* Lima: Instituto Nacional de Cultura.

Sheets, Payson, Christine Dixon, Monica Guerra, and Adam Blanford. 2011. "Manioc Cultivation at Ceren, El Salvador: Occasional Kitchen Garden Plant or Staple Crop?" *Ancient Mesoamerica* 22 (01):1–11.

Sheets, Payson, David Lentz, Dolores Piperno, John Jones, Christine Dixon, George Maloof, et al. 2012. "Ancient Manioc Agriculture South of the Ceren Village, El Salvador." *Latin American Antiquity* 23 (3):259–281.

Sinopoli, Carla M. 1991. *Approaches to Archaeological Ceramics.* New York: Plenum Press.

Slack, Charles R., and Marshall D. Hatch. 1967. "Comparative Studies on the Activity of Carboxylases and Other Enzymes in Relation to the New Pathway of Photosynthetic Carbon Dioxide Fixation in Tropical Grasses." *Biochemical Journal* 103 (3):660–5.

Smalley, John, and Michael Blake. 2003. "Sweet Beginnings: Stalk Sugar and the Domestication of Maize." *Current Anthropology* 44 (5):675–703.

Smith, Adam. 1812 [1776]. *An Inquiry into the Nature and Causes of the Wealth of Nations.* 3 vols. London: Cadell and Davies.

Smith, Bruce D. 1992. *Rivers of Change.* Washington, DC: Smithsonian Institution Press.

———. 1994. "The Origins of Agriculture in the Americas." *Evolutionary Anthropology: Issues, News, and Reviews* 3 (5):174–184.

———. 1995. *The Emergence of Agriculture.* New York: Scientific American Library.

———. 2001. "Documenting Plant Domestication: The Consilience of Biological and Archaeological Approaches." *Proceedings of the National Academy of Sciences, USA* 98 (4):1324–1326.

———. 2005. "Reassessing Coxcatlan Cave and the Early History of Domesticated Plants in Mesoamerica." *Proceedings of the National Academy of Sciences, USA* 102 (27):9438–9445.

———. 2006. "Documenting Domesticated Plants in the Archaeological Record." In *Documenting Domestication: New Genetic and Archaeological Paradigms,* edited by M.A. Zeder, D.G. Bradley, E. Emshwiller, and B.D. Smith, pp. 15–24. Berkeley: University of California Press.

Snow, David. 1990. "Tener Comal y Metate: Protohistoric Rio Grande Maize Use and Diet." In *Perspectives on Southwestern Prehistory,* edited by P.E. Minnis and C.L. Redman, pp. 289–300. Boulder: Westview Press.

Spencer, Charles S. 2010. "Territorial Expansion and Primary State Formation." *Proceeding of the National Academy of Sciences, USA* 107 (16):7119–7126.

Spiess, Arthur, and Leon Cranmer. 2001. "Native American Occupations at Pemaquid: Review and Recent Results." *Maine Archaeological Society Bulletin* 41:1–25.

Staller, John E. 2003. "An Examination of the Palaeobotanical and Chronological Evidence for an Early Introduction of Maize (*Zea mays* L.) into South America: A Response to Pearsall." *Journal of Archaeological Science* 30 (3):373–380.

———. 2006. "The Social, Symbolic, and Economic Significance of *Zea mays* L. In the Late Horizon Period." In *Histories of Maize: Multidisciplinary Approaches to the Prehistory, Biogeography, Domestication, and Evolution of Maize,* edited by J.E. Staller, R.H. Tykot, and B.F. Benz, pp. 449–467. Amsterdam: Academic Press.

———. 2010. *Maize Cobs and Cultures: History of* Zea mays L. Berlin: Springer-Verlag.

Staller, John E., and Robert G. Thompson. 2002. "A Multidisciplinary Approach to Understanding the Initial Introduction of Maize into Coastal Ecuador." *Journal of Archaeological Science* 29 (1):33–50.

Stothert, Karen E. 1985. "The Preceramic Las Vegas Culture of Coastal Ecuador." *American Antiquity* 50 (3):613–637.

Struever, Stuart. 1968. "Flotation Techniques for the Recovery of Small-Scale Archaeological Remains." *American Antiquity* 33:353–362.

Taube, Karl A. 1995. "The Olmec Maize God: The Face of Corn in Formative Mesoamerica." *RES: Anthroplogy and Aesthetics* 29/30:39–81.

———. 2000. "Lightning Celts and Corn Fetishes: The Formative Olmec and the Development of Maize Symbolism in Mesoamerica and the American Southwest." In *Olmec Art and Archaeology: Social Complexity in the Formative Period,* edited by J.E. Clark and M.E. Pye, pp. 297–337. Washington, DC: National Gallery of Art.

———. 2004. *Olmec Art at Dumbarton Oaks.* Washington, DC: Dumbarton Oaks Research Library and Collection.

———. 2010. "Gateways to Another World: The Symbolism of Supernatural Passageways in the Art and Ritual of Mesoamerica and the American Southwest." In *Painting the Cosmos: Metaphor and Worldview in Images from the Southwest Pueblos and Mexico,* edited by K.A. Hays-Gilpin and P. Schaafsma, pp. 73–120. Flagstaff: Museum of Northern Arizona.

Taylor, R.E., and Ofer Bar-Yosef. 2014. *Radiocarbon Dating: An Archaeological Perspective.* Walnut Creek, CA: Left Coast Press.

Tedlock, Dennis. 1985. *Popol Vuh: The Definitive Edition of the Mayan Book of the Dawn of Life and the Glories of Gods and Kings.* New York: Simon and Schuster.

Trigger, Bruce. 1989. *A History of Archaeological Thought.* Cambridge, UK: Cambridge University Press.

Tuxill, John, Luis Arias Reyes, Luis Latournerie Moreno, Vidal Cob Uicab, and Devra I. Jarvis. 2010. "All Maize Is Not Equal: Maize Variety Choices and Mayan Foodways in Rural Yucatan, Mexico." In *Pre-Columbian Foodways: Interdiscipilanary Approaches to Food, Culture, and Markets in Ancient Mesoamerica,* edited by J.E. Staller and M. Carrasco, pp. 467–486. New York: Springer.

Tykot, Robert H. 2006. "Isotope Analyses and the Histories of Maize." In *Histories of Maize: Multidisciplinary Approaches to the Prehistory, Biogeography, Domestication, and Evolution of Maize,* edited by J.E. Staller, R.H. Tykot, and B.F. Benz, pp. 131–142. Amsterdam: Academic Press.

Tykot, Robert H., Richard L. Burger, and Nikolaas J. van der Merwe. 2006. "The Importance of Maize in Initial Period and Early Horizon Peru." In *Histories of Maize: Multidisciplinary Approaches to the Prehistory, Biogeography, Domestication, and Evolution of Maize,* edited by J.E. Staller, R.H. Tykot, and B.F. Benz, pp. 187–197. Amsterdam: Academic Press.

Tykot, Robert H., and John E. Staller. 2002. "The Importance of Early Maize Agriculture in Coastal Ecuador: New Data from La Emerenciana." *Current Anthropology* 43 (4):666–677.

Tylor, Edward B. 1865. *Researches into the Early History of Mankind and the Development of Civilization.* London: John Murray.

UNC-Herbarium. 2015. Plant Information Center. Chapel Hill: University of North Carolina. www.ibiblio.org/pic/botanical_dictionary.htm, accessed March 4, 2015.

UNESCO. 2010. "Traditional Mexican Cuisine—Ancestral, Ongoing Community Culture, the Michoacán Paradigm." Paris: UNESCO. www.unesco.org/culture /ich/index.php?lg=en&pg=00011&RL=00400, accessed May 31, 2012.

Valdez, Francisco. 2008. "Inter-Zonal Relationships in Ecuador." In *Handbook of South American Archaeology,* edited by H. Silverman and W.H. Isbell, pp. 865–888. New York: Springer.

———. 2013. *Primeras Sociedades de la alta Amazonía: La cultura Mayo Chinchipe-Marañón.* Quito, Ecuador: Instituto Nacional de Patrimonio Cultural and Institut de Recherche pour le Développement.

van der Merwe, Nikolaas J., and John C. Vogel. 1978. "13C Content of Human Collagen as a Measure of Prehistoric Diet in Woodland North America." *Nature* 276:815–816.

van Heerwaarden, Joost, John Doebley, William H. Briggs, Jeffrey C. Glaubitz, Major M. Goodman, Jose de Jesus, et al. 2011. "Genetic Signals of Origin, Spread, and Introgression in a Large Sample of Maize Landraces." *Proceedings of the National Academy of Sciences, USA* 108 (3):1088–1092.

Vogel, John C., and Nikolaas J. van der Merwe. 1977. "Isotopic Evidence for Early Maize Cultivation in New York State." *American Antiquity* 42:238–242.

Vogt, Evon Z. 1969. *Zinacantan: A Maya Community in the Highlands of Chiapas.* Cambridge, MA: Belknap Press of Harvard University Press.

———. 1976. *Tortillas for the Gods: A Symbolic Analysis of Zinacanteco Rituals.* Cambridge, MA: Harvard University Press.

Voorhies, Barbara, George H. Michaels, and George M. Riser. 1991. "An Ancient Shrimp Fishery in South Coastal Mexico." *National Geographic Research and Exploration* 7 (1):20–35.

vPlants. 2009. The vPlants Project. vPlants: A Virtual Herbarium of the Chicago Region. www.vplants.org/plants/glossary/index.html, accessed July 14, 2012.

Wainwright, Milton. 2001. "Microbiology before Pasteur." *Microbiology Today* 28 (Feb):19–21.

Wang, Huai, Tina Nussbaum-Wagler, Bailin Li, Qiong Zhao, Yves Vigouroux, Marianna Faller, et al. 2005. "The Origin of the Naked Grains of Maize." *Nature* 436:714–719.

Wang, R. L., T. Ueda, and J. Messing. 1998. "Characterization of the Maize Prolamin Box-Binding Factor-1 (*Pbf-1*) and Its Role in the Developmental Regulation of the Zein Multigene Family." *Gene* 223:321–332.

Wellhausen, E. J., L. M. Roberts, E. Hernández X., and P. C. Mangelsdorf. 1952. *Races of Maize in Mexico.* Cambridge, MA: Bussey Institution of Harvard University.

Wenk, Gary. 2010. *Your Brain on Food: How Chemicals Control Your Thoughts and Feelings.* New York: Oxford University Press.

White, Christine D. 1997. "Diet at Lamanai and Pacbitun: Implications for the Ecological Model of Maya Collapse." In *Bones of the Maya: Studies of Ancient Skeletons,* edited by S. Whittington and D. Reed, pp. 171–181. Washington, DC: Smithsonian Institution Press.

White, Christine D., Paul F. Healy, and Henry P. Schwarcz. 1993. "Intensive Agriculture, Social Status, and Maya Diet at Pacbitun, Belize." *Journal of Anthropological Research* 49 (4):347–375.

White, Christine D., Fred J. Longstaffe, and Henry P. Schwarcz. 2006. "Social Directions in the Isotopic Anthropology of Maize in the Maya Region." In *Histories of Maize: Multidisciplinary Approaches to the Prehistory, Biogeography, Domestication, and Evolution of Maize,* edited by J. E. Staller, R. H. Tykot, and B. F. Benz, pp. 143–159. Amsterdam: Academic Press.

White, Christine D., David M. Pendergast, Fred J. Longstaffe, and Kimberley R. Law. 2001. "Social Complexity and Food Systems at Altun Ha, Belize: The Isotopic Evidence." *Latin American Antiquity* 12 (4):371–393.

Whitehead, Neil L., Michael J. Heckenberger, and George Simon. 2010. "Materializing the Past among the Lokono (Arawak) of the Berbice River, Guyana." *Antropológica* 54 (114):87–127.

Wilding, L.P. 1967. "Radiocarbon Dating of Biogenetic Opal." *Science* 156 (3771):66–67.

Wilkes, H. Garrison. 1967. *Teosinte: The Closest Relative of Maize.* Cambridge, MA: Bussey Institution of Harvard University.

Willey, Gordon R. 1966. *An Introduction to American Archaeology.* Vol. 1: *North and Middle America.* Englewood Cliffs, NJ: Prentice-Hall.

———. 1971. *An Introduction to American Archaeology.* Vol. 2: *South America.* Englewood Cliffs, NJ: Prentice-Hall.

Willis, Justin. 2002. *Potent Brews: A Social History of Alcohol in East Africa, 1850–1999.* Athens: Ohio University Press.

Wills, David M., Clinton J. Whipple, Shohei Takuno, Lisa E. Kursel, Laura M. Shannon, Jeffrey Ross-Ibarra, et al. 2013. "From Many, One: Genetic Control of Prolificacy during Maize Domestication." *PLoS Genetics* 9 (6):e1003604.

Wills, Wirt H. 1988. *Early Prehistoric Agriculture in the American Southwest.* Santa Fe: School of American Research.

Wilson, David J. 1981. "Of Maize and Men: A Critique of the Maritime Hypothesis of State Origins on the Coast of Peru." *American Anthropologist* 83 (1):93–120.

Wilson, Peter J. 1988. *The Domestication of the Human Species.* New Haven: Yale University Press.

Wright, Stephen I., and Brandon S. Gaut. 2005. "Molecular Population Genetics and the Search for Adaptive Evolution in Plants." *Molecular Biology and Evolution* 22:506–519.

Zarrillo, Sonia. 2012. "Human Adaptation, Food Production, and Cultural Interaction During the Formative Period in Highland Ecuador." PhD diss., University of Calgary.

Zhao, Da-Zhong, Guan-Fang Wang, Brooke Speal, and Hong Ma. 2002. "The Excess Microsporocytes1 Gene Encodes a Putative Leucine-Rich Repeat Receptor Protein Kinase That Controls Somatic and Reproductive Cell Fates in the Arabidopsis Anther." *Genes & Development* 16:2021–2031.

Index

Milton Keynes UK
Ingram Content Group UK Ltd.
UKHW022113270524
443238UK00005B/210